THEORY
OF
MATRICES

SAM PERLIS
Professor Emeritus of Mathematics
Purdue University

D0171219

DOVER PUBLICATIONS, INC.
New York

Published in Canada by General Publishing Company, Ltd., 30 Lesmill Road, Don Mills, Toronto, Ontario.

Published in the United Kingdom by Constable and Company, Ltd., 3 The Lanchesters, 162–164 Fulham Palace Road, London W6 9ER.

This Dover edition, first published in 1991, is an unabridged and unaltered republication of the 1958 third printing of the work first published by the Addison-Wesley Publishing Company, Inc., Reading, Mass., in 1952.

Manufactured in the United States of America
Dover Publications, Inc., 31 East 2nd Street, Mineola, N.Y. 11501

Library of Congress Cataloging-in-Publication Data

Perlis, Sam, 1913–
    Theory of matrices / Sam Perlis. — Dover ed.
        p.   cm.
    Reprint. Originally published: Reading, Mass. : Addison-Wesley Pub. Co., 1952.
    Includes bibliographical references and index.
    ISBN 0-486-66810-X (pbk.)
    1. Matrices   I. Title.
QA188.P42   1991
512.9'434—dc20
                                               91-19037
                                                 CIP

TO IDA AND ESTHER

PREFACE

The remarkable growth of applications of matrix theory is well known. Although other branches of higher mathematics may be applied more intensively and perhaps by more people, few branches have been applied to so many diversified fields as has the theory of matrices. Education, psychology, chemistry, physics; electrical, mechanical, and aeronautical engineering; statistics and economics — all are enriched by this theory. Within mathematics itself the applications of matrices to other mathematical disciplines are widespread.

These facts are reflected in the composition and magnitude of university classes in matrices. The students specializing in mathematics are sometimes outnumbered as much as ten to one. Such conditions present a teaching problem, inasmuch as students from "foreign fields" rarely have adequate maturity for advanced mathematical reasoning.

For a subject so much in demand, the texts on matrices are surprisingly few. Of the texts published in this country within the last decade or two some have approached the teaching problem cited above by the process of excision. Much of the meat of the subject is cut away and a skeleton is exposed. This fare, well seasoned with numerical work, is then offered as the sole diet. Other texts use a method of concentration. The theory is carried almost no further than the subject of inverses or in some cases the Cayley-Hamilton theorem, and then attention is concentrated on a specialized field of application such as electrical networks.

Although both of these approaches have obvious value and fill a need, I feel that still another is sorely needed. My planning of this book has been guided by several observations. First, the difficulties which confront students of engineering, psychology, chemistry, and other subjects, when they take a course in matrix theory are often not so much the result of deficient knowledge of the subject matter of earlier courses as of a lack of mathematical maturity. In very little of their experience have definition, theorem, and proof played so prominent a role. The second observation influencing me is that students from "foreign fields" are generally no less intelligent than those who specialize in mathematics (although some would say they are less fortunate!). Third, I feel that even if the student is primarily concerned with applications to a narrow technical field, successful employment of this subject depends ultimately on facing and mastering the main ideas.

In relation to this third point we may consider the current interest in efficient methods for inverting matrices of large order. An instructor might easily be tempted to devote a large portion of a beginning course to this topic, but there are reasons for regarding such a temptation with disfavor. In the first place, research organizations will usually present the new employee with a handbook outlining the methods of inversion they are using. In the second place, the methods themselves are undergoing development, so that any method learned in class would very likely be subject to rapid obsolescence. There is certainly no objection to an advanced or second course devoted to current techniques, but the beginner would do much better to devote his time to the basic ideas. This, at any rate, is one of the personal viewpoints which have guided my writing.

With these ideas in mind I have planned a book which offers a more than ample diet for mathematics majors taking a first course of one semester in the theory of matrices. I have attempted to make this diet palatable to students with meager backgrounds by exercising great care and patience in writing proofs, discussions, and examples to illustrate definitions and theorems. This coddling, however, is supplemented by activities leading the student to stand on his own feet. There is a generous supply of exercises including adequate amounts of numerical work, but bearing heavily on simple theoretical questions. Occasional queries are inserted in the text proper to focus attention on a fine point or to stimulate review of earlier topics. Where parallel theories occur, one is worked out in detail and the other merely outlined, details of proof being left for the student to work out and perhaps to present as a classroom report. In all these ways the student gains familiarity with the significance of definitions and theorems, and gains confidence in himself.

The main theme of the book is the establishment of the well-known canonical forms. Rank, nonsingularity, and inverses are introduced in connection with the development of canonical matrices under the relation of equivalence, and without the intervention of determinants. Congruence and Hermitian congruence are the next major topics. Among the several methods of treating similarity,

I have chosen one which seems to me to promote the maximum enlightenment for the student. In Chapter 9 characteristic vectors and roots lead to the various diagonalization theories, culminating in the study of normal matrices.

If brevity were a consideration of overriding urgency, parallel theories would in some instances have been treated simultaneously, notably in the theories relating to Hermitian and real, symmetric matrices. I make no apology for separating the cases. The student gains much by seeing that simple ideas can with suitable modifications be extended to further cases, and by fashioning his own proofs from the models presented in the simpler cases. This type of presentation is particularly featured in Chapter 9, where the subject matter of diagonalization progresses from real, symmetric matrices, to Hermitian matrices, thence to normal matrices.

Certain exercises throughout the book are distinguished by a star. Contrary to common usage, this designation does not indicate that the problem is difficult. Starred problems are those to which a reference is made later in the book, or which state a property of general interest. The student should at least read every starred exercise.

Numerous supplementary topics are treated in appendixes located at the ends of various chapters. Thus they are available near the point at which they become relevant but are not intrusions upon the orderly development of the main theorems.

The arrangement of chapters lends itself well to a variety of courses. For an elementary course in which considerable class time is devoted to problem solving, the first five chapters will suffice. The same students could cover substantially all of the book in a one-year course. For students with a better background, experience has shown that the first eight chapters can be covered successfully in a semester course of about fifty lectures. If, further, a knowledge of determinants and polynomials as presented in Chapters 4 and 6 is a prerequisite for the course, omission of these chapters will then permit covering in one semester most of the material to the end of Chapter 9.

Lafayette, Indiana   SAM PERLIS
April, 1952

CONTENTS

CHAPTER 1

INTRODUCTORY CONCEPTS

1-1 Simultaneous linear equations. If we consider a system of linear equations, such as

(1)
$$2x + 3y - z = 7,$$
$$-10x - 2y + 5z = 5,$$

it is soon apparent that all investigations of the system may be carried out, perhaps with greater efficiency, by working only with the array of coefficients and the array of constant terms:

(2)
$$\begin{bmatrix} 2 & 3 & -1 \\ -10 & -2 & 5 \end{bmatrix}, \quad \begin{bmatrix} 7 \\ 5 \end{bmatrix}.$$

Although the exploration of this idea is deferred until Chapter 3, we have at once the concept of a rectangular array of numbers. The first array in (2) has two rows and three columns; the second has two rows and one column.

Any rectangular array of numbers is called a *matrix* or a *rectangular matrix*, and if the array has n rows and s columns it is called an $n \times s$ (read "n by s") matrix. Two matrices are *equal* if they are precisely the same arrays (though they may differ notationally, just as x^2 differs from xx). A convenient notation for a general $n \times s$ matrix is found in the use of double subscripts:

(3)
$$\begin{bmatrix} a_{11} & a_{12} & \ldots & a_{1s} \\ a_{21} & a_{22} & \ldots & a_{2s} \\ \cdot & \cdot & & \cdot \\ \cdot & \cdot & & \cdot \\ \cdot & \cdot & & \cdot \\ a_{n1} & a_{n2} & \ldots & a_{ns} \end{bmatrix}.$$

The s numbers at the top of this array constitute the first row of the matrix. The rows of a matrix are numbered from the top down, and the columns from left to right, so that the first subscript on a quantity a_{ij} indicates the row in which it belongs, and the second subscript indicates the column. The quantities a_{ij} are called the *quantities* or *elements* or *coordinates* or *entries* of the matrix. Some-

1

times we refer to a_{ij} as the element in the (i,j) *place*. If the dimensions n and s of the matrix (3) are kept in mind, it suffices to use the brief notation (a_{ij}) in place of (3). Various symbols of enclosure are in common use:

$$(a_{ij}), [a_{ij}], \|a_{ij}\|.$$

We shall generally use parentheses for matrices having only one row, and brackets for all other matrices. Moreover, it is common to write $A = (a_{ij})$ and use the single letter A to denote a matrix (3).

One may be tempted to place the symbols x, y, and z above the appropriate columns in (2) as reminders of the system (1). The matrix notation requires, as we shall see in the next section, that we write these "reminders" in the following way:

$$(4) \qquad \begin{bmatrix} 2 & 3 & -1 \\ -10 & -2 & 5 \end{bmatrix} \begin{bmatrix} x \\ y \\ z \end{bmatrix}, \quad \begin{bmatrix} 7 \\ 5 \end{bmatrix}.$$

Regardless of where the labels x, y, and z are placed, the left side of the first equation in (1) may be reconstructed from the first row of the 2×3 matrix in (2) by multiplying each of the elements $2, 3, -1$ in this row by x, y, and z, respectively, then adding:

$$2x + 3y + (-1)z.$$

Similarly the left side of the second equation is obtained from the second row of the 2×3 matrix in (2) by multiplying the -10 by x, the -2 by y, the 5 by z, and adding. In terms of the notation (4), this suggests a kind of row-by-column multiplication. This suggestion will arise with greater force in the next section.

1-2 Substitutions and matrix multiplication. Let us consider a general system like (1):

$$(5) \qquad \begin{aligned} l_1 &= a_{11}x + a_{12}y + a_{13}z, \\ l_2 &= a_{21}x + a_{22}y + a_{23}z. \end{aligned}$$

Such a system of equations can sometimes be simplified by means of a substitution, say,

$$(6) \qquad \begin{aligned} x &= b_{11}u + b_{12}v, \\ y &= b_{21}u + b_{22}v, \\ z &= b_{31}u + b_{32}v. \end{aligned}$$

Here the letters u and v denote new unknowns or variables, and the letters b_{ij} denote constants. After the substitution is made, the system appears as

$$l_1 = c_{11}u + c_{12}v,$$
$$l_2 = c_{21}u + c_{22}v,$$

where the 2×2 matrix of coefficients $C = (c_{ij})$ is

(7)
$$\begin{bmatrix} a_{11}b_{11} + a_{12}b_{21} + a_{13}b_{31} & a_{11}b_{12} + a_{12}b_{22} + a_{13}b_{32} \\ a_{21}b_{11} + a_{22}b_{21} + a_{23}b_{31} & a_{21}b_{12} + a_{22}b_{22} + a_{23}b_{32} \end{bmatrix}.$$

For discussion purposes, let A and B denote the coefficient matrices in (5) and (6):

$$A = \begin{bmatrix} a_{11} & a_{12} & a_{13} \\ a_{21} & a_{22} & a_{23} \end{bmatrix}, \quad B = \begin{bmatrix} b_{11} & b_{12} \\ b_{21} & b_{22} \\ b_{31} & b_{32} \end{bmatrix}.$$

Then the element

$$c_{11} = a_{11}b_{11} + a_{12}b_{21} + a_{13}b_{31}$$

in C has been obtained from A and B by "multiplying" the first row of A by the first column of B, that is, the first row of A and the first column of B each have first elements (a_{11} and b_{11}), second elements (a_{12} and b_{21}), and third elements (a_{13} and b_{31}); each pair is multiplied and the results are added. This gives the formula above for c_{11}. In similar fashion, c_{ik} is obtained by multiplying row i of A by column k of B in the manner just described.

This illustration suggests the following definition. Let $A = (a_{ij})$ be an $n \times s$ matrix, and $B = (b_{jk})$ be an $s \times t$ matrix. Then the *product* AB is defined as the matrix

$$C = (c_{ik}), \quad c_{ik} = \sum_{j=1}^{s} a_{ij}b_{jk}.$$

The wisdom of this definition is borne out if we generalize the illustration above. Instead of (5) consider a system of n linear equations in s unknowns $x_1, \ldots, x_s$:

(8)
$$l_i = \sum_{j=1}^{s} a_{ij}x_j. \qquad (i = 1, \ldots, n)$$

This compact notation for a system of n equations in s unknowns displays clearly its $n \times s$ coefficient matrix $A = (a_{ij})$. Suppose that in (8) we wish to make the substitution

(9)
$$x_j = \sum_{k=1}^{t} b_{jk}y_k \qquad (j = 1, \ldots, s)$$

with coefficient matrix $B = (b_{jk})$. The result is the system

$$(10) \qquad l_i = \sum_{j=1}^{s} a_{ij} \left(\sum_{k=1}^{t} b_{jk} y_k \right),$$

or

$$(11) \qquad l_i = \sum_{k=1}^{t} \left(\sum_{j=1}^{s} a_{ij} b_{jk} \right) y_k,$$

where in each case $i = 1, \ldots, n$.

The passage from (10) to (11) is described as a change in the order of summation. We sum first for k in (10), but first for j in (11). If we accept for the moment the validity of this change (see Section 1–11 for proof) we conclude that the result (11) of the substitution is a system of n linear equations in t unknowns $y_1, \ldots, y_t$ with coefficient matrix

$$C = (c_{ik}), \quad c_{ik} = \sum_{j=1}^{s} a_{ij} b_{jk}.$$

According to the definition above, $C = AB$. We have proved

THEOREM 1–1. *If a system of linear equations* (8) *with coefficient matrix A is subjected to a substitution* (9) *with coefficient matrix B, the resulting system* (11) *in new unknowns has coefficient matrix AB.*

Let X and L denote the $s \times 1$ and $n \times 1$ matrices

$$(12) \qquad X = \begin{bmatrix} x_1 \\ \cdot \\ \cdot \\ \cdot \\ x_s \end{bmatrix}, \quad L = \begin{bmatrix} l_1 \\ \cdot \\ \cdot \\ \cdot \\ l_n \end{bmatrix}.$$

Then the system (8) may be written very compactly by use of matrices:

$$(13) \qquad AX = L.$$

This formula (13) makes the system (8) appear as simple as a single linear equation in one unknown, $ax = b$. One of our goals is to develop an algebra for matrices which will permit the solution of the system (13) by methods analogous to, and theoretically as simple as, the solution of $ax = b$. To avoid space-consuming displays like (12) we introduce the notation

$$X = \text{col} (x_1, \ldots, x_s),$$
$$L = \text{col} (l_1, \ldots, l_n)$$

for the one-columned matrices (12). Commas are usually used to separate the elements of a matrix having only one row; otherwise there is danger of confusion with an ordinary product of numbers.

The reader must bear in mind two points concerning matrix products AB. If A is $n \times r$ and B is $s \times t$, the product AB is defined only when $r = s$; the product AB is undefined when, for example, A is 2×3 and B is 5×2. The second point is that when A is $n \times s$ and B is $s \times t$, the product AB is $n \times t$: the number of rows in AB is the number of rows in A (the left factor), and the number of columns in AB is the number of columns in B.

A matrix is said to be *square* if its number of rows is equal to its number of columns. This number is sometimes called the *order* of the matrix.

EXERCISES

1. Given the matrices

$$A = \begin{bmatrix} 1 & 2 & 3 \\ 0 & -1 & 1 \\ 2 & 3 & 0 \end{bmatrix}, \quad B = \begin{bmatrix} 3 & 1 & 0 \\ 1 & -1 & 2 \\ 0 & 2 & 1 \end{bmatrix},$$

verify the following products:

$$AB = \begin{bmatrix} 5 & 5 & 7 \\ -1 & 3 & -1 \\ 9 & -1 & 6 \end{bmatrix}, \quad BA = \begin{bmatrix} 3 & 5 & 10 \\ 5 & 9 & 2 \\ 2 & 1 & 2 \end{bmatrix}.$$

2. In the first system of equations below make the substitution given by the second system of equations (a) by direct substitution and (b) by use of matrices:

$$\begin{cases} 2x + 3y = 1, \\ x + 2y = 5, \end{cases} \quad \begin{cases} x = 2u - 3v, \\ y = -u + 2v. \end{cases}$$

(c) Solve the new system of equations for u and v, and thus find the values of x and y satisfying the original system.

3. Given the matrices

$$A = \begin{bmatrix} 1 & 2 \\ 3 & 4 \end{bmatrix}, \quad B = \begin{bmatrix} 3 & 0 \\ -1 & 4 \end{bmatrix}, \quad C = (2, 3), \quad D = \begin{bmatrix} 1 \\ -1 \end{bmatrix},$$

calculate the products AB, BA, CD, DC, AD, CA.

1-3 Transpose and associativity. Quadratic as well as linear polynomials may sometimes be associated with matrices. For example, consider

$$\begin{aligned} f(x, y) &= 3x^2 + 4xy - y^2 \\ &= 3x^2 + 2xy + 2yx - y^2 \\ &= x(3x + 2y) + y(2x - y). \end{aligned}$$

Now $f(x, y)$ may be regarded as a 1×1 matrix which is equal to the matrix product

$$f(x, y) = (x, y)\begin{bmatrix} 3x + 2y \\ 2x - y \end{bmatrix},$$

(14) $$f(x, y) = (x, y)\begin{bmatrix} 3 & 2 \\ 2 & -1 \end{bmatrix}\begin{bmatrix} x \\ y \end{bmatrix}.$$

This subject will be treated carefully in Chapter 5. For the moment (14) will be used to introduce several topics.

The matrix col (x, y) appears in (14) together with (x, y). Each of these two is called the transpose of the other. More generally, if $A = (a_{ij})$ is an $n \times s$ matrix, its *transpose* is defined as the $s \times n$ matrix

$$B = (b_{pq}), \quad b_{ji} = a_{ij}$$

for all i and j. We may think of constructing B from A, column by column, by rotating the first row of A and using it as the first column of B, the second row as the second column, and so on.

The only elements of A that do not change positions during this process are those whose row and column subscripts are equal: a_{11}, a_{22}, ... These elements constitute the *diagonal* of A, and individually are called *diagonal elements* of A.

The transpose of a matrix A will be denoted by the symbol A'.

THEOREM 1–2. *The transpose of a product AB is the product of the transposes in reverse order:* $(AB)' = B'A'$.

Let $A = (a_{ij})$ be $n \times s$ and $B = (b_{jk})$ be $s \times t$, so that the element in the (i,k) place (ith row and kth column) of AB is

(15) $$d_{ik} = \sum_{j=1}^{s} a_{ij}b_{jk}.$$

This is the element in the (k,i) place of $(AB)'$. The element in the (k,i) place of $B'A'$ is the product of the kth row of B' by the ith column of A'. But these are $(b_{1k}, b_{2k}, \ldots, b_{sk})$ and col $(a_{i1}, a_{i2}, \ldots, a_{is})$, and their product is (15).

The wording of this theorem suggests that if PQ is a matrix product such that QP also is defined, these two products may differ. This is in fact the case; it is comparatively rare to have $PQ = QP$. For example,

$$P = \begin{bmatrix} 1 & 2 \\ 3 & 4 \end{bmatrix}, \quad Q = \begin{bmatrix} 0 & 1 \\ 0 & 0 \end{bmatrix}, \quad PQ = \begin{bmatrix} 0 & 1 \\ 0 & 3 \end{bmatrix}, \quad QP = \begin{bmatrix} 3 & 4 \\ 0 & 0 \end{bmatrix}.$$

In elementary algebra the property that $ab = ba$ for all choices of a and b is known as the *commutative law*. As a law, commutativity fails to be true for matrix multiplication. For special pairs P and Q, however, it may be true that $PQ = QP$, and when this is true we say that P *commutes with* Q.

One other issue arises from (14) by virtue of the fact that it is a product of three matrices. With ordinary numbers a triple product $(ab)c$ may be computed also as $a(bc)$. The equality between these products is known as the *associative law*.

THEOREM 1–3. *Matrix multiplication is associative:*

$$P(QR) = (PQ)R.$$

The proof is a direct computation. Let

$$P = (p_{ij}), \quad Q = (q_{jk}), \quad R = (r_{kl})$$

so that the element in the (i,k) place in $PQ = (d_{ik})$ is

$$d_{ik} = \sum_j p_{ij}q_{jk},$$

and in the (i,l) place of $(PQ)R$ is

(16) $$\sum_k d_{ik}r_{kl} = \sum_k \sum_j p_{ij}q_{jk}r_{kl}.$$

The element in the (i,l) place of $P(QR)$ may be computed similarly. First, the (j,l) place of QR is occupied by the element

$$c_{jl} = \sum_k q_{jk}r_{kl},$$

whence the element in the (i,l) place of $P(QR)$ is

$$\sum_j p_{ij}c_{jl} = \sum_j \sum_k p_{ij}q_{jk}r_{kl}.$$

This is the same as (16), since the two differ only in the order of summation. This completes the proof and shows that a triple product PQR is unambiguous even if parentheses are omitted.

The proof of Theorem 1–1 actually proved a special case of the associative law. We are now in a position to repeat that proof in very brief form. The system of linear equations may be written [see (13)]

(17) $$AX = L$$

and the substitution (9) may be written $X = BY$. Then substitution in (17) gives

$$A(BY) = L = (AB)Y,$$

by Theorem 1–3. Hence the new system has AB as coefficient matrix.

EXERCISES

1. Given the matrices

$$A = \begin{bmatrix} 1 & 0 & 3 \\ 2 & -1 & 1 \end{bmatrix}, \quad B = \begin{bmatrix} 3 & 4 & 1 \\ 0 & -1 & 5 \\ 1 & 2 & -2 \end{bmatrix}, \quad C = \begin{bmatrix} 2 \\ -1 \\ 4 \end{bmatrix}.$$

Calculate the following products:

(a) $(AB)'$, $B'A'$; (b) $(AC)'$, $C'A'$; (c) $(BC)'$, $C'B'$.

2. If A, B, and C are the matrices of Exercise 1, test the associative law for the product ABC.

3. Find all matrices B obeying the equation

$$\begin{bmatrix} 0 & 1 \\ 0 & 2 \end{bmatrix} B = \begin{bmatrix} 0 & 0 \\ 0 & 0 \end{bmatrix}.$$

4. Find all matrices B obeying the equation

$$\begin{bmatrix} 0 & 1 \\ 0 & 2 \end{bmatrix} B = \begin{bmatrix} 0 & 0 & 1 \\ 0 & 0 & 2 \end{bmatrix}.$$

5. Find all matrices B which commute with

$$A = \begin{bmatrix} 0 & 1 \\ 0 & 2 \end{bmatrix}.$$

★6.* Use Theorem 1–3 to prove the associative law for products $ABCD$ of four matrices. That is, prove the equality of all five of the following products: $[(AB)C]D$, $[A(BC)]D$, $A[(BC)D]$, $A[B(CD)]$, $(AB)(CD)$. The associative law is true (can you prove it?) for products of any number of matrices: no matter how parentheses are inserted so that only two matrices are to be multiplied in each step, the same product matrix is always obtained.

★7. If P commutes with Q, show that P and Q are square and of the same size.

★8. If P commutes with Q, show that P' commutes with Q'.

---

* A star preceding the number of a problem indicates that the problem sets forth a property with which the student should become familiar. Since there are frequent references to such properties, the student should read, and preferably work, all starred problems.

1-4 Diagonal matrices. A matrix A is called *diagonal* if it is square,* say $n \times n$, and if its elements off the diagonal are all zero:

(18)
$$A = \begin{bmatrix} a_1 & 0 & \dots & 0 \\ 0 & a_2 & \dots & 0 \\ \cdot & \cdot & & \cdot \\ \cdot & \cdot & & \cdot \\ \cdot & \cdot & & \cdot \\ 0 & 0 & \dots & a_n \end{bmatrix}.$$

To save space we shall use the notation

(19)
$$A = \mathrm{diag}\,(a_1, \dots, a_n)$$

to denote the matrix above. It is a simple exercise to verify that a product AB is obtained from B by multiplying the ith row of B by a_i, $i = 1, \dots, n$. In like manner BA is computed by multiplying the jth column of B by a_j, $j = 1, \dots, n$.

If the elements a_i in (19) are all equal to a common element c,

(20)
$$A = \mathrm{diag}\,(c, c, \dots, c),$$

A is called a *scalar matrix*. Then AB is computed by multiplying every element of B by c, and the same is true of BA (if the dimensions are such that BA is defined). For this reason a product AB or BA, where A is the scalar matrix above, is denoted by cB, and this is called a *scalar product*, or a process of *scalar multiplication*. Also, the elements a_{ij} of matrices $A = (a_{ij})$ are called *scalars* for the purpose of sharp contrast with matrices themselves.

A scalar matrix whose diagonal elements are equal to unity is called an *identity matrix* and is denoted by I. When it is necessary to indicate the number n of rows and columns in I, the symbol I_n is used in place of I. Hence $I_n B = B = B I_s$ for every $n \times s$ matrix B. Also the scalar product cB is equal to both of the matrix products, $(cI_n)B = B(cI_s) = cB$.

EXERCISES

1. If $A = \mathrm{diag}\,(a_1, \dots, a_n)$ and $B = \mathrm{diag}\,(b_1, \dots, b_n)$, prove that $AB = BA$.

2. If $A = cI_n$, prove that A commutes with every $n \times n$ matrix.

3. If B is a general 3×3 matrix (b_{ij}) and $A = \mathrm{diag}\,(a_1, a_2, a_3)$, compute AB and BA.

---

* Occasionally, as in Chapter 7, this term is also applied to nonsquare matrices (a_{ij}) with $a_{ij} = 0$ for $i \neq j$.

1-5 Addition. The *sum* of two $n \times s$ matrices $A = (a_{ij})$ and $B = (b_{ij})$ is defined in the natural way by adding similarly placed elements:

$$A + B = C = (c_{ij}), \quad c_{ij} = a_{ij} + b_{ij}.$$

All familiar number systems possess a quantity z such that $x + z = x$ for every quantity x in the system, namely $z = 0$. For $n \times s$ matrices there is a quantity similar in this respect to 0, namely the $n \times s$ matrix whose elements are all 0. It is called the $n \times s$ *zero matrix*. Each zero matrix will usually be denoted by the symbol 0; the context will suffice to distinguish it from the scalar 0 and to make clear the dimensions of the zero matrix.

Another analogy between the properties of the scalar 0 and of zero matrices is the obvious fact that

$$0B = 0, \quad B0 = 0.$$

In these equations, however, if B is $n \times s$ and the first zero matrix is $m \times n$, the second zero matrix must be $m \times s$; the third zero is $s \times t$ for some integer t, and the last zero is $n \times t$.

The property in elementary algebra that $a(b + c) = ab + ac$ is known as the *distributive law*. Its matrix analogue is true.

THEOREM 1-4. *Matrix multiplication is distributive:*

$$A(B + C) = AB + AC, \quad (B + C)D = BD + CD.$$

To prove the first of these, let $A = (a_{ij})$, $B = (b_{jk})$, and $C = (c_{jk})$, so that $B + C = (b_{jk} + c_{jk})$. The element in the (i,k) place of $A(B + C)$ is then

$$\sum_j a_{ij}(b_{jk} + c_{jk}) = \sum_j a_{ij}b_{jk} + \sum_j a_{ij}c_{jk},$$

which is the sum of the elements in the (i,k) places of AB and AC. The second distributive law is proved similarly.

The concept of addition may also be used in discussing a product AB. Suppose, for example, that

$$A = \begin{bmatrix} 2 & 3 \\ 4 & 5 \end{bmatrix}, \quad B = \begin{bmatrix} b_1 & e_1 \\ b_2 & e_2 \end{bmatrix}.$$

Then if $AB = C = (c_{ij})$, we have

$$C = \begin{bmatrix} 2b_1 + 3b_2 & 2e_1 + 3e_2 \\ 4b_1 + 5b_2 & 4e_1 + 5e_2 \end{bmatrix}.$$

If we let B_i and C_i denote the ith rows of B and C, respectively, we may condense this formula for C:

$$C = \begin{bmatrix} C_1 \\ C_2 \end{bmatrix} = \begin{bmatrix} 2B_1 + 3B_2 \\ 4B_1 + 5B_2 \end{bmatrix}.$$

Here $2B_1$ and $3B_2$ are the scalar products

$$2B_1 = (2b_1, \quad 2e_1),$$
$$3B_2 = (3b_2, \quad 3e_2),$$

so that

$$2B_1 + 3B_2 = (2b_1 + 3b_2, \quad 2e_1 + 3e_2)$$
$$= C_1.$$

In similar fashion

$$4B_1 + 5B_2 = C_2.$$

More generally, if $B = (b_{jk})$ is $s \times t$, its rows

$$B_j = (b_{j1}, b_{j2}, \ldots, b_{jt})$$

are $1 \times t$ matrices B_j. If $A = (a_{ij})$, the (i,k) place in $C = AB = (c_{ik})$ is occupied by

$$(21) \qquad c_{ik} = a_{i1}b_{1k} + a_{i2}b_{2k} + \cdots + a_{is}b_{sk}.$$

The ith row C_i of C is

$$C_i = (c_{i1}, c_{i2}, \ldots, c_{it}),$$

and we may conclude from (21) that

$$(22) \qquad C_i = a_{i1}B_1 + a_{i2}B_2 + \cdots + a_{is}B_s.$$

Here each term $a_{ij}B_j$ is a scalar product, and the two sides of (22) are $1 \times t$ matrices whose kth elements are equal by (21). An expression like that on the right side of (22) is called a *linear* combination of the rows $B_1, \ldots, B_s$ with coefficients $a_{i1}, \ldots, a_{is}$.

The reader may describe the product $C = AB$ of 2×2 matrices above in terms of columns of A and C. Generally, if A is partitioned into its columns

$$A^{(j)} = \text{col } (a_{1j}, a_{2j}, \ldots, a_{nj}),$$

the kth column of $C = AB$ is found to be the matrix sum

$$(23) \qquad C^{(k)} = A^{(1)}b_{1k} + A^{(2)}b_{2k} + \cdots + A^{(s)}b_{sk};$$

that is, the kth column of AB is a linear combination of the columns of A with coefficients taken from the kth column of B.

THEOREM 1–5. *Each row of AB is a linear combination of the rows of B, and each column of AB is a linear combination of the columns of A.*

It is important to remember not only the idea in Theorem 1–5 but also the manner in which the linear combinations are formed. If each row B_j of B is regarded as a single object, B takes on the appearance of a column matrix,

$$B = \text{col } (B_1, \ldots, B_s).$$

Then (22) shows that AB may be computed by multiplying each row of A by the "one column" of B. Equation (23) yields a similar statement wherein the columns of A are regarded as single objects, and A is a row of these objects.

EXERCISES

1. If

$$P = \begin{bmatrix} 2 & 1 \\ -1 & 3 \end{bmatrix}, \quad Q = \begin{bmatrix} 0 & 1 \\ 2 & 5 \end{bmatrix}, \quad R = \begin{bmatrix} 2 & 3 \\ 4 & 0 \end{bmatrix},$$

compute $2P + 3Q + R$.

2. In Exercise 1 find the value of c if

$$P + 2Q + cR = \begin{bmatrix} 0 & 0 \\ -1 & 13 \end{bmatrix}.$$

3. If the ith and jth rows of a matrix A are alike, prove that the ith and jth rows in any product AB must be alike.

4. If the 5th row of A is $(1, -1, 0, 0)$, describe the 5th row of AB in terms of rows of B.

5. If $A = \text{col } (1, -1, 0, 0)$, describe BA in terms of columns of B. Do the same if $A = \text{col } (1, 1, 1, 1)$.

6. Given that

$$B = \begin{bmatrix} 1 & 1 & 1 & 1 \\ 1 & 3 & 2 & 2 \\ 1 & -3 & -1 & -1 \\ 1 & -5 & -2 & -2 \end{bmatrix}, \quad A = \begin{bmatrix} 1 \\ 1 \\ 1 \\ c \end{bmatrix},$$

show that there is a number c (and find its value) such that $BA = 0$; such that $A'B = (4, -4, 0, 0)$.

7. If B is the matrix of Exercise 6 and $A = \text{col } (a_1, 1, 1, a_4)$, find a_1 and a_4 if $BA = \text{col } (0, 2, -4, -6)$.

8. Let $P, Q,$ and R be fixed $n \times n$ matrices. Prove that A commutes with every matrix $c_1P + c_2Q + c_3R$ if and only if A commutes with $P, Q,$ and R.

9. Prove that a 2×2 matrix A commutes with every 2×2 matrix if and only if A commutes with the four matrices

$$\begin{bmatrix} 1 & 0 \\ 0 & 0 \end{bmatrix}, \quad \begin{bmatrix} 0 & 1 \\ 0 & 0 \end{bmatrix}, \quad \begin{bmatrix} 0 & 0 \\ 1 & 0 \end{bmatrix}, \quad \begin{bmatrix} 0 & 0 \\ 0 & 1 \end{bmatrix}.$$

Find all such matrices A.

★10. Using a method suggested by Exercise 9, prove that an $n \times n$ matrix A commutes with every $n \times n$ matrix if and only if A is scalar.

★11. Prove the following rules for transposes:

$$(A + B)' = A' + B', \ (cA)' = cA'.$$

12. Find formulas for the following products:
   (a) $(A + B)(C + D)$;
   (b) $(A + B)(A + B)$;
   (c) $(A + B)(A - B)$.

13. Prove that $(A + B)(A - B) = A^2 - B^2$ if and only if A commutes with B.

14. The sum of the diagonal elements of a matrix A is called the *trace* of A and denoted by tr (A). Prove that tr $(A + B)$ = tr (A) + tr (B), and tr $(cA) = c \cdot$ tr (A).

1-6 Partitioned matrices. If any rows or columns, or both, of a matrix $A = (a_{ij})$ are deleted, the rectangular array that remains is called a *submatrix* of A. Each element a_{ij} is a submatrix obtained by deleting all rows except row i and all columns except column j. Also, if A has at least four rows and six columns,

$$B = \begin{bmatrix} a_{23} & a_{25} & a_{26} \\ a_{43} & a_{45} & a_{46} \end{bmatrix}$$

is the submatrix obtained by deleting all rows but the second and fourth, and all columns but the third, fifth, and sixth. Each row or column of a matrix is a submatrix.

The trick of multiplying matrices by partitioning them into rows or columns can be pursued further. Let

$$A = \begin{bmatrix} a_{11} & a_{12} & a_{13} & a_{14} & a_{15} \\ a_{21} & a_{22} & a_{23} & a_{24} & a_{25} \\ a_{31} & a_{32} & a_{33} & a_{34} & a_{35} \\ a_{41} & a_{42} & a_{43} & a_{44} & a_{45} \end{bmatrix} = (A_1, A_2),$$

where, as indicated, A_1 is the 4×2 submatrix made up of the first two columns of A, and A_2 is the 4×3 submatrix, whose columns are the last three columns of A. Let B be a 5×3 matrix so that the product AB is defined, and partition B into

$$B = \begin{bmatrix} b_{11} & b_{12} & b_{13} \\ b_{21} & b_{22} & b_{23} \\ \hline b_{31} & b_{32} & b_{33} \\ b_{41} & b_{42} & b_{43} \\ b_{51} & b_{52} & b_{53} \end{bmatrix} = \begin{bmatrix} B_1 \\ B_2 \end{bmatrix},$$

where B_1 and B_2 are the indicated submatrices of B. We shall verify that

$$(24) \qquad AB = (A_1, A_2)\begin{bmatrix} B_1 \\ B_2 \end{bmatrix} = A_1B_1 + A_2B_2,$$

so that AB is the sum of the matrix products A_1B_1 and A_2B_2. The partitionings of A and B have been chosen so that these products are defined: the numbers of columns of A_1 and rows of B_1 are both 2, and for A_2 and B_2 these numbers are both 3.

To prove (24), let C_i denote the ith row of $C = AB = (c_{ik})$, and R_i the ith row of B, so that

$$(25) \qquad C_i = (a_{i1}R_1 + a_{i2}R_2) + (a_{i3}R_3 + a_{i4}R_4 + a_{i5}R_5)$$

by (22). The first parenthetical expression in (25) is the ith row of A_1B_1 and the second is the ith row of A_2B_2, whence C_i (the ith row of AB) is the sum of these ith rows. This statement amounts to (24). It is clear that this proof works for A and B of any sizes such that AB is defined. Thus

THEOREM 1–6. *Let A and B be matrices which are partitioned as follows:*

$$A = (A_1, A_2, \ldots, A_r), \quad B = \begin{bmatrix} B_1 \\ \cdot \\ \cdot \\ \cdot \\ B_r \end{bmatrix}.$$

Then $AB = A_1B_1 + \cdots + A_rB_r$, provided only that every product A_iB_i is defined.

While only the case $r = 2$ was discussed above, the generalization is immediate. A further generalization is commonly employed. Since the product $C = AB$ may be computed a row at a time, it makes no difference if $A = (A_1, \ldots, A_r)$ is further partitioned by a horizontal line drawn through it at any place:

(26)
$$A = \begin{bmatrix} A_{11} & A_{12} & \ldots & A_{1r} \\ A_{21} & A_{22} & \ldots & A_{2r} \end{bmatrix}.$$

It can then be seen that

(27)
$$AB = \begin{bmatrix} A_{11}B_1 + \cdots + A_{1r}B_r \\ A_{21}B_1 + \cdots + A_{2r}B_r \end{bmatrix}.$$

Suppose that the upper layer in (26) includes only the first s rows in A. Then (27) is proved in exactly the same way as Theorem 1–6, except that the cases $i \leqq s$ and $i > s$ are considered separately. Moreover, it is clear that any number of "horizontal separators" may be passed through A,

(28)
$$A = \begin{bmatrix} A_{11} & \ldots & A_{1r} \\ \cdot & & \cdot \\ \cdot & & \cdot \\ \cdot & & \cdot \\ A_{q1} & \ldots & A_{qr} \end{bmatrix}, \quad B = \begin{bmatrix} B_{11} & \ldots & B_{1t} \\ \cdot & & \cdot \\ \cdot & & \cdot \\ \cdot & & \cdot \\ B_{r1} & \ldots & B_{rt} \end{bmatrix},$$

and any number of "vertical separators" through B. The result, which the reader can easily verify, is stated as follows:

THEOREM 1–7. *Let A and B be matrices such that AB is defined and A and B are partitioned as in (28). If the products $A_{ij}B_{jk}$ are defined for all values of i, j, and k, AB may be computed in terms of submatrices as follows:*

$$AB = \begin{bmatrix} C_{11} & \ldots & C_{1t} \\ \cdot & & \cdot \\ \cdot & & \cdot \\ \cdot & & \cdot \\ C_{q1} & \ldots & C_{qt} \end{bmatrix},$$

$$C_{ij} = A_{i1}B_{1j} + A_{i2}B_{2j} + \cdots + A_{ir}B_{rj}.$$

Thus AB may be computed with the "row-by-column" rule of multiplication applied to (28) as if the submatrices there were ordinary elements. Submatrices appearing in partitionings like (28) are frequently called "blocks," and the multiplication described in Theorem 1–7 is called *block multiplication*. A block is thus a submatrix of A selected from a succession of adjacent rows and adjacent columns.

Block multiplication is particularly handy if $q = r = t$ in (28) and all blocks off the diagonal are zero. Then by analogy with (19) we may write

(29) $A = \operatorname{diag}(A_1, \ldots, A_r), \quad B = \operatorname{diag}(B_1, \ldots, B_r),$

and prove at once from Theorem 1–7 that

$$AB = \text{diag } (A_1B_1, \ldots, A_rB_r).$$

Here again the blocks behave like ordinary elements.   But we cannot conclude now that $AB = BA$ unless we know that $A_iB_i = B_iA_i$, $i = 1, \ldots, r$.

A matrix like A in (29) is called the *direct sum* of $A_1, A_2, \ldots, A_r$. This terminology will be employed later.

EXERCISES

1. If

$$A = \begin{bmatrix} 2 & 3 & 4 \\ 1 & 2 & 7 \end{bmatrix}, \quad B = \begin{bmatrix} -1 & 3 & 2 \\ 1 & 0 & -2 \\ 2 & 1 & 1 \end{bmatrix},$$

calculate AB first by linear combinations of rows, second by linear combinations of columns.

2. Calculate the product AB in Exercise 1 by writing $A = (A_1, A_2)$, where A_1 is 2×2, and

(a) $B = \begin{bmatrix} -1 & 3 & 2 \\ 1 & 0 & -2 \\ \hline 2 & 1 & 1 \end{bmatrix};$

(b) $B = \begin{bmatrix} -1 & 3 & 2 \\ 1 & 0 & -2 \\ \hline 2 & 1 & 1 \end{bmatrix};$

(c) $B = \begin{bmatrix} -1 & 3 & 2 \\ 1 & 0 & -2 \\ 2 & 1 & 1 \end{bmatrix}.$

1–7 Fields of scalars.   Thus far nothing has been said about the nature of the scalars, the elements composing our matrices.   Some readers may have assumed that they are arbitrary real numbers, others that they are complex numbers.   Whatever they are, our usage of them indicates that they constitute a number system in which addition and multiplication are defined and are subject to the usual laws of elementary algebra.   As a matter of fact, the applications of matrices require that the scalar number system be the reals for some purposes and the complexes for others.   The modern mathematician requires many other number systems.   In order to avoid repeated proofs for the various systems, we consider a generalization including all of them, a type of system called a field.

Each of the familiar number systems such as the reals or the complexes is a collection $\mathfrak{F}$ of objects together with two operations called addition and multiplication. Addition associates with each pair, u and v, of elements of $\mathfrak{F}$ a unique quantity of $\mathfrak{F}$ called the sum of u and v and denoted by $u + v$. Sums have the following properties for all u, v, and w in $\mathfrak{F}$:

A1. $(u + v) + w = u + (v + w)$.

A2. $u + v = v + u$.

A3. $\mathfrak{F}$ has a unique element 0 such that $u + 0 = u$.

A4. For each u in $\mathfrak{F}$ there is a unique v (denoted by $-u$) in $\mathfrak{F}$ such that $u + v = 0$.

Multiplication associates with each pair, u and v, of elements of $\mathfrak{F}$ a unique quantity of $\mathfrak{F}$ called the product of u and v, denoted by uv, and having the following properties for all u, v, and w in $\mathfrak{F}$:

M1. $(uv)w = u(vw)$.

M2. $uv = vu$.

M3. $\mathfrak{F}$ has a unique element 1, different from 0, such that $1u = u = u1$.

M4. For each $u \neq 0$ and in $\mathfrak{F}$, there is a unique v (denoted by u^{-1}) in $\mathfrak{F}$ such that $uv = 1 = vu$.

Connecting the two operations are the distributive laws, valid for all u, v, and w:

D.
$$u(v + w) = uv + uw,$$
$$(v + w)u = vu + wu.$$

Any collection $\mathfrak{F}$ of objects on which sums and products are so defined that they lie in $\mathfrak{F}$ and obey the laws $A1$–4, $M1$–4, and D, is called a *field*.

Subtraction and division are defined in a field by the equations

$$u - v = u + (-v),$$
$$u/v = u(v^{-1}). \qquad (v \neq 0)$$

The four elementary operations (also called *rational* operations) are thus available in any field. One can prove from the postulates $A1$–4, $M1$–4, and D that the manipulations elaborated in the elementary texts on algebra are valid in any field. Since it would divert attention too far from matrices, these proofs will not be given here; the reader who is interested in developing the fundamentals of field theory is referred to numbers 3, 4, 10, and 17 in the References.

Polynomials

$$a_n x^n + a_{n-1} x^{n-1} + \cdots + a_1 x + a_0$$

with coefficients a_i in a field $\mathfrak{F}$ are said to be polynomials *over* $\mathfrak{F}$. These, too, are subject to the usual manipulations of algebra.

The totality of real numbers and the totality of complex numbers afford two common examples of fields. The sum and product of two real numbers are again real numbers, and the properties $A1$–4, $M1$–4, and D are well known to be true for these numbers; and similarly for complex numbers. Some less familiar examples of fields are set forth in the next section and its exercises. The important point is that the concept of a field is an abstraction covering many number systems, in every one of which quantities may be added, subtracted, multiplied, and divided according to the rules with which the reader has long been familiar.

A matrix is said to be *over* $\mathfrak{F}$ if its elements belong to the field $\mathfrak{F}$, and a theorem is true over $\mathfrak{F}$ if it is true when $\mathfrak{F}$ is the field of scalars for all matrices concerned. As far as possible the theorems in this book will be stated and proved with a general field as the field of scalars. In some parts of the theory the results are not known over a general scalar field, or they take different forms for different fields. Suitable restrictions on the field will be stated in these cases.

1–8 Sets. Any collection of entities is known as a *set* or a *class* The students in a particular room, the books in a library, the trees in a forest, the letters on this page, all are examples of sets. The objects belonging to a set $\mathcal{S}$ are called *members* of $\mathcal{S}$, and are said to *belong to* $\mathcal{S}$ or *lie in* $\mathcal{S}$. If $\mathcal{S}$ and $\mathcal{J}$ are sets and every member of $\mathcal{S}$ is also a member of $\mathcal{J}$, we say that $\mathcal{S}$ is *contained in* $\mathcal{J}$ (notationally: $\mathcal{S} \leq \mathcal{J}$), or $\mathcal{J}$ *contains* $\mathcal{S}$ (notationally: $\mathcal{J} \geq \mathcal{S}$), or that $\mathcal{S}$ is a *subset* (or *subclass*) of $\mathcal{J}$. It is important to note that $\mathcal{J} \leq \mathcal{J}$, that is, every set is a subset of itself.

The set $\mathfrak{R}$ of all real numbers is thus a subset of the set $\mathfrak{C}$ of all complex numbers. Since $\mathfrak{R}$ and $\mathfrak{C}$ are fields, and $\mathfrak{R}$ is contained in $\mathfrak{C}$, $\mathfrak{R}$ is also called a *subfield* of $\mathfrak{C}$. A matrix over $\mathfrak{R}$ is called *real*, and a matrix over $\mathfrak{C}$ is called *complex*. The notations $\mathfrak{R}$ and $\mathfrak{C}$ will be reserved throughout for the fields of all real and all complex numbers, respectively. The student who is interested only in the practical applications of matrices may prefer to think only of real or complex matrices when, as in most of the following work, we discuss matrices over a general field.

If $\mathfrak{F}$ is any field, the sum of two quantities (or members) of $\mathfrak{F}$ is again in $\mathfrak{F}$. This fact is expressed by the statement that $\mathfrak{F}$ is *closed under addition.* Similarly $\mathfrak{F}$ is closed under multiplication, subtraction, and division. (Division by zero is tacitly excluded from consideration.) Thus we may say that a field $\mathfrak{F}$ is closed under the four elementary (or rational) operations. This, in fact, is the major idea motivating the concept of a field.

It can be verified that if $\mathfrak{F}_0$ is a subset of a field $\mathfrak{F}$ such that (a) $\mathfrak{F}_0$ contains at least two members and (b) $\mathfrak{F}_0$ is closed under the four rational operations already defined in $\mathfrak{F}$, then $\mathfrak{F}_0$ is a field, hence a subfield of $\mathfrak{F}$. If we accept this criterion, we can find many interesting subfields of $\mathfrak{C}$. One important example is the field of all rational numbers, which simply are all ordinary quotients of integers.

A less familiar example of a field is the set $\mathfrak{F}$ of all quotients

$$\frac{P(x)}{Q(x)}$$

of polynomials with real numbers as coefficients. This set $\mathfrak{F}$ is closed under addition, since

$$\frac{P_1(x)}{Q_1(x)} + \frac{P_2(x)}{Q_2(x)} = \frac{P_1(x)Q_2(x) + P_2(x)Q_1(x)}{Q_1(x)Q_2(x)}.$$

Both numerator and denominator on the right are polynomials over $\mathfrak{R}$; we have therefore shown that a sum of two members of $\mathfrak{F}$ is always in $\mathfrak{F}$. $\mathfrak{F}$ is also closed under multiplication:

$$\frac{P_1(x)}{Q_1(x)} \cdot \frac{P_2(x)}{Q_2(x)} = \frac{P_1(x)P_2(x)}{Q_1(x)Q_2(x)}.$$

These closure properties are the crucial ones in this example. The reader can verify each of the properties $A1$–4, $M1$–4, and D.

EXERCISES

1. Verify that each of the following subsets of $\mathfrak{C}$ forms a subfield. (Hint: Show that each subset is closed under the four rational operations.) (a) The rational numbers. (b) All quantities $a + b\sqrt{2}$, a and b rational. (c) All quantities $a + b\sqrt[3]{2} + c\sqrt[3]{4}$, a, b, and c rational. (d) All quantities $a + b\sqrt{-1}$, a and b rational.

2. Show that the rational number field has no subfield other than itself.

1–9 Square matrices as a number system. For each field $\mathfrak{F}$ let $\mathfrak{F}_n$ denote the totality of $n \times n$ matrices over $\mathfrak{F}$. Since $\mathfrak{F}_n$ is a collection of objects on which addition and multiplication are defined, it is interesting to see how many of the properties defining a field **are**

valid for $\mathfrak{F}_n$, and to note carefully those field properties which are not true in $\mathfrak{F}_n$.

Since matrix addition is carried out elementwise with correspondingly placed elements, properties $A1$ and $A2$ are clearly valid for $\mathfrak{F}_n$ in consequence of their validity in $\mathfrak{F}$. For $A3$ the $n \times n$ zero matrix clearly meets all requirements. If $U = (u_{ij})$, take $V = (v_{ij})$, $v_{ij} = -u_{ij}$. Then $U + V = 0$, and this V is the only matrix whose sum with U is 0. This verifies $A4$, and all of the addition postulates for a field are valid for $\mathfrak{F}_n$.

As for the remaining postulates, we have already verified D and $M1$. The identity matrix I_n serves as the 1 of postulate $M3$. Both $M2$ and $M4$, however, are invalid. For example,

$$A = \begin{bmatrix} 1 & 2 \\ 3 & 4 \end{bmatrix}, \qquad B = \begin{bmatrix} 0 & 1 \\ 0 & 0 \end{bmatrix},$$

$$AB = \begin{bmatrix} 0 & 1 \\ 0 & 3 \end{bmatrix}, \qquad BA = \begin{bmatrix} 3 & 4 \\ 0 & 0 \end{bmatrix},$$

so that $M2$ is not universally satisfied in $\mathfrak{F}_2$. Also, B above is a nonzero quantity of $\mathfrak{F}_2$, but there is no matrix

$$C = \begin{bmatrix} c_{11} & c_{12} \\ c_{21} & c_{22} \end{bmatrix}$$

such that $BC = I$ or $CB = I$; therefore $M4$ fails.

Any set of objects on which addition and multiplication are so defined as to obey postulates $A1$–4, $M1$, and D is called a *ring*. If $M2$ is true the ring is called *commutative*, and if $M3$ is true the ring is said to *have a unity*. Hence

THEOREM 1–8. $\mathfrak{F}_n$ *forms a noncommutative ring with a unity.*

One of the most familiar examples of a ring is the totality of ordinary integers (positive, negative, and zero). While $\mathfrak{F}_n$ thus has much in common with the integers, it also has some startling differences. Noncommutativity, as observed above, is one such difference. Another is the presence in $\mathfrak{F}_n$ of matrices A and B, both nonzero, such that $AB = 0$. Such a matrix A is called a *left divisor of zero*, and B is called a *right divisor of zero*. Example:

$$A = \begin{bmatrix} -2 & 1 \\ 2 & -1 \end{bmatrix}, \quad B = \begin{bmatrix} 2 & 3 \\ 4 & 6 \end{bmatrix}, \quad AB = 0.$$

We shall not refer to Theorem 1–8 or continue the discussion of rings, but there will be occasion to treat matrices which are divisors of zero. The main purpose of this section is to underline both the similarities and the distinctions between the algebra of square matrices and elementary algebra or, as we may call it, the algebra of scalars.

EXERCISES

1. If A and B are matrices such that $AB = 0$ and $B \neq 0$, prove that there is no matrix C such that $CA = I$.

2. If A is the real matrix

$$A = \begin{bmatrix} -2 & 1 \\ 2 & -1 \end{bmatrix},$$

prove that the matrices

$$B = \begin{bmatrix} b_1 & b_2 \\ 2b_1 & 2b_2 \end{bmatrix}$$

are the only 2×2 matrices B such that $AB = 0$.

3. If A is the matrix of Exercise 2, find all 2×2 matrices C such that $CA = 0$. Then use the result of Exercise 2 to find all matrices B such that $AB = BA = 0$.

★4. Let A be a square matrix over $\mathfrak{F}$. Prove that A commutes with every matrix of the type $c_r A^r + c_{r-1} A^{r-1} + \cdots + c_1 A + c_0 I$, where the c_i are in $\mathfrak{F}$.

1–10 Matrices of one row or one column. In Section 1–2 we saw that a system of simultaneous linear equations can be written in the form $AX = L$, where $X = \text{col } (x_1, \ldots, x_s)$ denotes a column matrix. The system $AX = L$ may be regarded in the following light: A operates to convert the column matrix X into the column matrix L.

Column matrices appeared again in Theorem 1–5, where it was pointed out that each column of a matrix product is a linear combination of the columns of the left factor; a similar result for rows was observed.

All of these results indicate that matrices having a single row or a single column may deserve special attention. The next chapter will be devoted to such matrices.

APPENDIX

1–11 Order of summation. If $C = (c_{jk})$ is an $n \times s$ matrix, the sum of the elements of C is given by the formula

$$S = \sum_{j=1}^{n} \sum_{k=1}^{s} c_{jk}.$$

This formula indicates that we sum first for k, keeping j fixed; that is, we find the sum of the elements in row j. There are n such sums, one for each row, and these are added when we sum for j as required by the second summation symbol. The same sum S is obtained if we first add all the elements in each column, then add these column sums. But this process is indicated by the double summation,

$$S = \sum_{k=1}^{s} \sum_{j=1}^{n} c_{jk}.$$

The two formulas thus found for S prove that

(30)
$$\sum_{j=1}^{n} \sum_{k=1}^{s} c_{jk} = \sum_{k=1}^{s} \sum_{j=1}^{n} c_{jk}.$$

That is, in any finite double sum the order of summation may be reversed without affecting the value of the sum.

In formulas (10) and (11) the subscript i remains fixed throughout the summations, so that we may safely use the notation

$$c_{jk} = a_{ij}b_{jk}y_k.$$

The equality of (10) and (11) then is assured by (30) above. In the proof of Theorem 1–3 we must know that

(31)
$$\sum_{k} \sum_{j} p_{ij}q_{jk}r_{kl} = \sum_{j} \sum_{k} p_{ij}q_{jk}r_{kl}.$$

Since i and l are fixed, we let

$$c_{jk} = p_{ij}q_{jk}r_{kl},$$

so that the desired equality (31) is a consequence of (30).

CHAPTER 2

VECTOR SPACES

2–1 Introduction. The representation of points of the plane by means of ordered pairs (x, y) of real numbers, or points of space by ordered triples (x, y, z), is a familiar device. These ideas may readily be associated with the vectors* employed in physics to represent' forces and other concepts. Suppose, for example, that two forces f_1 and f_2 acting on a particle are represented by vectors emanating from the origin and terminating at the points whose rectangular coordinates are (x_1, y_1, z_1), and (x_2, y_2, z_2), respectively. We may then use the notation,

$$(1) \qquad f_1 = (x_1, y_1, z_1), \quad f_2 = (x_2, y_2, z_2).$$

The resultant force on the particle is denoted by $f_1 + f_2$ and is known to be represented by the vector (emanating from the origin) which is the diagonal of the parallelogram having the two given vectors as sides. One may verify that the coordinates of the end points of $f_1 + f_2$ are the sums of corresponding coordinates of f_1 and f_2:

$$f_1 + f_2 = (x_1 + x_2, y_1 + y_2, z_1 + z_2).$$

This property is known as the *parallelogram law* for addition of vectors.

Another familiar operation on the vectors of physics is that of multiplying by a real number. Thus kf_1 is a vector having the same direction as f_1 if k is positive, or the opposite direction if k is negative, and having length equal to $|k|$ times that of f_1. One then finds that

$$(2) \qquad kf_1 = (kx_1, ky_1, kz_1).$$

If k is 0, kf_1 is defined as the zero vector, which is merely one point, the origin. Thus (2) is valid even for $k = 0$.

2–2 Definitions. The concepts of the previous section may be generalized by considering ordered n-tuples

$$(3) \qquad \xi = (x_1, \ldots, x_n)$$

---

* Arrows whose directions indicate the direction of, say, the forces, and whose magnitudes are proportional to the magnitudes of the forces.

of quantities x_i lying in a field $\mathfrak{F}$. Each n-tuple (3) is called a *vector* with n *coordinates* or *components* x_i. The totality $V_n(\mathfrak{F})$ of all vectors (3), for a fixed integer n, is called the n-*dimensional vector space over* $\mathfrak{F}$.

If $\eta = (y_1, \ldots, y_n)$ is another vector belonging to $V_n(\mathfrak{F})$, the *sum* $\xi + \eta$ is defined as the vector

$$\xi + \eta = (x_1 + y_1, \ldots, x_n + y_n),$$

and for each scalar k (quantity in $\mathfrak{F}$), the *scalar product* $k\xi$ is defined as

$$k\xi = (kx_1, \ldots, kx_n).$$

It will be recognized that $V_n(\mathfrak{F})$ is the totality of $1 \times n$ matrices over $\mathfrak{F}$, and that the addition and scalar multiplication defined here are simply the matrix operations defined in Chapter 1. However, it is of no importance that ξ is written as a row. It may be written as a column, and $V_n(\mathfrak{F})$ may be regarded as the totality of $n \times 1$ matrices over $\mathfrak{F}$. Thus ordered n-tuples will be regarded here as objects which may be represented as rows or columns, whichever may be convenient at the moment. In a later section of this chapter we will consider still another way in which vectors are sometimes written.

That vector of $V_n(\mathfrak{F})$ whose components are all zero is called the *zero vector* and is denoted by 0. Although this is the same symbol used for the zero scalar, no confusion can arise. In the equation $\xi + \eta = 0$, ξ and η being vectors, the 0 certainly cannot be a scalar, and in the expression 0ξ the zero cannot be a vector, since (for the present) we do not multiply vectors. Lower case Greek letters will always denote vectors. Other symbols also will be used for vectors in various special situations, but a small Greek letter will never denote anything but a vector.

For every vector $\xi = (c_1, \ldots, c_n)$ in $V_n(\mathfrak{F})$, there is a vector η such that

$$\xi + \eta = 0,$$

namely the vector $\eta = (-c_1, \ldots, -c_n)$. Clearly this is the only vector η whose sum with ξ is the zero vector. This vector η, uniquely determined by ξ, is called the *negative* of ξ and is denoted by $-\xi$. Since $\eta = (-1)\xi$ the definition of $-\xi$ may be written

$$-\xi = (-1)\xi.$$

Subtraction of vectors is defined by the equation

$$\eta - \xi = \eta + (-1)\xi.$$

It is easy to verify the following properties, valid for all vectors ξ, $\xi_1, \ldots, \xi_t$ in $V_n(\mathfrak{F})$ and all scalars $k, k_1, \ldots, k_t$:

$$(4) \qquad (\xi_1 + \xi_2) + \xi_3 = \xi_1 + (\xi_2 + \xi_3), \ \xi_1 + \xi_2 = \xi_2 + \xi_1,$$
$$(5) \qquad k(\xi_1 + \cdots + \xi_t) = k\xi_1 + \cdots + k\xi_t,$$
$$(6) \qquad (k_1 + \cdots + k_t)\xi = k_1\xi + \cdots + k_t\xi,$$
$$(7) \qquad (k_1 k_2)\xi = k_1(k_2\xi).$$

The properties (4) are merely special cases of the corresponding laws for general rectangular matrices.

These properties or rules permit us to solve simple vector equations or systems of equations very much like simple algebraic equations. If

$$\xi + 5\eta = (8, -5, 6),$$
$$\xi - \eta = (2, \ 1, 0),$$

then adding 5 times the second equation to the first gives

$$6\xi = (18, 0, 6), \quad \xi = (3, 0, 1).$$

whence

$$\eta = \xi - (2, 1, 0) = (1, -1, 1).$$

2-3 Subspaces. Let V be any collection of vectors belonging to $V_n(\mathfrak{F})$. Then V is *closed under addition* if V contains $\xi + \eta$ for every ξ and every η lying in V. Similarly, V is *closed under scalar multiplication* if the presence of ξ in V implies the presence of $k\xi$ in V for every scalar k.

DEFINITION 1. A subset V (containing at least one vector) of $V_n(\mathfrak{F})$ is called a *subspace* of $V_n(\mathfrak{F})$, or simply* a *vector space over* $\mathfrak{F}$, if V is closed under addition and scalar multiplication.

One obvious subspace of $V_n(\mathfrak{F})$ is $V_n(\mathfrak{F})$ itself. Another is the subset Z consisting only of the zero vector and called the *zero* or *null* subspace. To verify that Z is a subspace, observe that if ξ and η belong to Z, both must be 0, so that their sum $\xi + \eta = 0 + 0 = 0$ also belongs to Z, and every scalar product $k\xi = k \cdot 0 = 0$ belongs to Z.

Every vector space V contains the zero vector. For V contains some vector $\xi = (c_1, \ldots, c_n)$, hence contains the scalar product, $0\xi = (0, \ldots, 0) = 0$. Also, V contains the negative, $-\xi = (-1)\xi$, of each vector ξ in V.

---

* Strictly, the vector spaces defined here are called *finite* or *finite-dimensional*. These adjectives will be omitted, since no other type of vector space will be studied here.

If V contains vectors $\xi_1, \ldots, \xi_t$, it must also contain every vector

(8) $$\xi = k_1\xi_1 + \cdots + k_t\xi_t,$$

where $k_1, \ldots, k_t$ are arbitrary scalars. Any vector having the form (8) is called a linear combination of $\xi_1, \ldots, \xi_t$.

THEOREM 2–1. *Let* $\xi_1, \ldots, \xi_t$ *belong to* $V_n(\mathfrak{F})$. *Then the totality V of linear combinations* (8) *is a subspace.*

Let ξ in (8) and

$$\eta = h_1\xi_1 + \cdots + h_t\xi_t$$

be any two vectors belonging to V. Then

$$\xi + \eta = (k_1 + h_1)\xi_1 + \cdots + (k_t + h_t)\xi_t,$$
$$k\xi = (kk_1)\xi_1 + \cdots + (kk_t)\xi_t$$

are linear combinations of the ξ_i. Thus V is closed under addition and scalar multiplication, and hence is a vector space.

DEFINITION 2. A vector space V is said to be *spanned* by vectors $\xi_1, \ldots, \xi_t$ if (1) $\xi_1, \ldots, \xi_t$ lie in V and (2) every vector in V is a linear combination of $\xi_1, \ldots, \xi_t$.

Another way of stating this definition is that V coincides with the totality of linear combinations of $\xi_1, \ldots, \xi_t$. Thus in Theorem 2–1 V is spanned by $\xi_1, \ldots, \xi_t$.

Let each vector (x, y, z) in $V_3(\mathfrak{R})$ be interpreted as the point whose rectangular coordinates are x, y, z. Each vector (x, y) in $V_2(\mathfrak{R})$ may also be interpreted as a point. Then each nonzero subspace of $V_3(\mathfrak{R})$ or $V_2(\mathfrak{R})$ becomes a collection of points which may be shown to be $V_3(\mathfrak{R})$ or a line or plane through the origin. For example, the subspace spanned by $\xi = (1, 2)$ consists of all points $(x, 2x)$, hence is the line $y = 2x$, and this is the line determined by the origin and the point $(1, 2)$. The subspace of $V_3(\mathfrak{R})$ spanned by $(1, 2, 3)$ consists of all points $(x, 2x, 3x)$, hence is the locus of the equations $y = 2x$, $z = 3x$, known to be the line connecting the origin and $(1, 2, 3)$. Again, the subspace spanned by $\xi = (1, 0, 0)$ and $\eta = (0, 1, 0)$ consists of all points $(x, y, 0)$, and hence is the xy plane.

The xy plane is a subspace V of $V_3(\mathfrak{R})$ spanned by ξ and η above. However, V is also spanned by

$$\tfrac{1}{3}\xi = (\tfrac{1}{3}, 0, 0), \quad \tfrac{1}{2}\eta = (0, \tfrac{1}{2}, 0),$$

since V contains these vectors, and every vector $(x, y, 0)$ in V is expressible in the form

$$(x, y, 0) = 3x(\tfrac{1}{3}\xi) + 2y(\tfrac{1}{2}\eta).$$

Thus, a spanning set for a subspace V is by no means unique.

EXERCISES

1. Let (x_1, x_2, x_3) be an arbitrary vector in $V_3(\Re)$. Which of the following subsets are subspaces? (a) All vectors with $x_1 = x_2 = x_3$. (b) All vectors with $x_1 = 0$. (c) All vectors with x_1, x_2, and x_3 rational. (d) All vectors with $x_1 = 1$.

2. Show that all the vectors (x_1, x_2, x_3, x_4) in $V_4(\Re)$ which obey $x_4 - \dot{x}_3 = x_2 - x_1$ form a subspace V. Then show that V is spanned by $\xi_1 = (1, 0, 0, -1)$, $\xi_2 = (0, 1, 0, 1)$, $\xi_3 = (0, 0, 1, 1)$.

3. Let a_1, a_2, and a_3 be fixed real numbers. Show that all vectors (x_1, x_2, x_3, x_4) in $V_4(\Re)$ obeying $x_4 = a_1x_1 + a_2x_2 + a_3x_3$ form a subspace V. Then show that V is spanned by $\xi_1 = (1, 0, 0, a_1)$, $\xi_2 = (0, 1, 0, a_2)$, $\xi_3 = (0, 0, 1, a_3)$

4. Let S be a subset of $V_n(\mathfrak{F})$ containing at least one vector. Prove that S is a subspace if and only if S contains $a\xi + b\eta$ for all vectors ξ and η in S and all scalars a and b.

5. Show that a subspace V of $V_3(\Re)$ coincides with $V_3(\Re)$ if and only if V contains the vectors $(1, 0, 0)$, and $(0, 1, 0)$, and $(0, 0, 1)$. Then determine which of the following sets span $V_3(\Re)$:

(a) $\xi_1 = (-1, 2, 3)$, $\xi_2 = (0, 1, 2)$, $\xi_3 = (3, 2, 1)$;

(b) $\eta_1 = (0, 0, 2)$, $\eta_2 = (2, 2, 0)$, $\eta_3 = (0, 2, 2)$;

(c) $\zeta_1 = (3, 3, 1)$, $\zeta_2 = (1, 1, 0)$, $\zeta_3 = (0, 0, 1)$.

6. Consider the subspaces of $V_3(\Re)$ spanned by each of the following sets of vectors. Determine which of these subspaces coincide with the subspace spanned by the vectors in (a).

(a) $\xi_1 = (1, 1, 1)$, $\xi_2 = (0, 1, 2)$, $\xi_3 = (1, 0, -1)$;

(b) $\eta_1 = (2, 1, 0)$, $\eta_2 = (2, 0, -2)$;

(c) $\zeta_1 = (1, 2, 3)$, $\zeta_2 = (1, 3, 5)$;

(d) ζ_1, ζ_2, and $\zeta_3 = (1, 2, 4)$.

2–4 Linear independence and bases. If $\xi_1, \ldots, \xi_t$ lie in a vector space V over $\mathfrak{F}$, they are called *linearly dependent* if there are scalars $k_1, \ldots, k_t$, not all zero, such that

$$(9) \qquad\qquad k_1\xi_1 + \cdots + k_t\xi_t = 0.$$

If no such scalars exist, the set $\xi_1, \ldots, \xi_t$, is called *linearly independent* (briefly, *independent*).

The vectors

$$\xi_1 = (1, 2, 1), \quad \xi_2 = (0, 1, 0), \quad \xi_3 = (2, 0, 2)$$

in $V_3(\Re)$ are linearly dependent, since

$$2\xi_1 - 4\xi_2 - \xi_3 = 0.$$

But if

$$h\xi_2 + k\xi_3 = 0,$$

then

$$h\xi_2 + k\xi_3 = (2k, h, 2k) = (0, 0, 0),$$

so that $h = k = 0$. Thus ξ_2 and ξ_3 are linearly independent, while ξ_1 is a linear combination of them:

$$\xi_1 = 2\xi_2 + \tfrac{1}{2}\xi_3.$$

The geometric interpretation of these facts is that ξ_2 and ξ_3 span a subspace which is a plane through the origin, and ξ_1 lies in this plane.

DEFINITION 3. If vectors $\xi_1, \ldots, \xi_t$ in a vector space V are linearly independent and span V, they are said to form a *basis* of V.

A basis for $V_n(\mathfrak{F})$ is readily found:

$$u_1 = (1, 0, \ldots, 0),$$
$$u_2 = (0, 1, 0, \ldots, 0),$$
$$\cdot \quad \cdot \quad \cdot \quad \cdot \quad \cdot$$
$$u_n = (0, 0, \ldots, 0, 1).$$

These are obviously independent, and every vector $\xi = (x_1, \ldots, x_n)$ is a linear combination of them:

$$\xi = x_1 u_1 + \cdots + x_n u_n.$$

These vectors $u_1, \ldots, u_n$ are called *unit vectors*, and the notations u_i will be reserved exclusively for them. There are, however, many other bases — for example, $u_1, \ldots, u_{n-1}$, and $\xi = (1, 1, \ldots, 1)$.

THEOREM 2-2. *If $\xi_1, \ldots, \xi_t$ form a basis for V, every vector in V is expressible uniquely as a linear combination of $\xi_1, \ldots, \xi_t$.*

Every ξ in V is a linear combination of the ξ_i, since they span V. If

$$\xi = a_1\xi_1 + \cdots + a_t\xi_t = b_1\xi_1 + \cdots + b_t\xi_t,$$

then

$$(a_1 - b_1)\xi_1 + \cdots + (a_t - b_t)\xi_t = 0,$$

so that

$$a_i - b_i = 0 \qquad (i = 1, \ldots, t)$$

because of the independence of the ξ_i. This shows that every $a_i = b_i$, and thus establishes the uniqueness.

We shall see that every vector space V has a basis, and shall determine what aspects of uniqueness a basis has. The next two lemmas will play fundamental roles in solving these problems.

LEMMA 2-1. Let $\xi_1, \ldots, \xi_t$ be nonzero vectors belonging to a vector space V over $\mathfrak{F}$. Then the set $\xi_1, \ldots, \xi_t$ is dependent if and only if some ξ_m $(1 < m \leq t)$ is a linear combination of $\xi_1, \ldots, \xi_{m-1}$. In this case the subspace spanned by the ξ_i $(i = 1, \ldots, t)$ is also spanned by those ξ_i with $i \neq m$.

If the ξ_i are dependent, an equation like (9) holds with at least one $k_i \neq 0$. If only one k_i were nonzero we should have $k_i\xi_i = 0$, $\xi_i = (x_1, \ldots, x_n) \neq 0$ by hypothesis, so that some $x_j \neq 0$, $k_i x_j = 0$, whence $k_i = 0$, a contradiction. Hence at least two of the k_i are nonzero. The last nonzero coefficient k_m in (9) then has subscript $m > 1$, and we can solve for ξ_m:

$$(10) \qquad \xi_m = (-k_1/k_m)\xi_1 + \cdots + (-k_{m-1}/k_m)\xi_{m-1}.$$

The converse is evident. In any linear combination $a_1\xi_1 + \cdots + a_t\xi_t$, the term $a_m\xi_m$ may by (10) be replaced by a sum of terms involving only $\xi_1, \ldots, \xi_{m-1}$. This completes the proof.

Now suppose that V is spanned by a finite set $\xi_1, \ldots, \xi_t$ of nonzero vectors. If these vectors are dependent, select the first ξ_m which is a linear combination of the ξ_i preceding it, and delete it from the list $\xi_1, \ldots, \xi_t$. The remaining ξ_i still span V according to Lemma 2-1. If this set of $t - 1$ vectors is dependent, a repetition of the process leads to a set of $t - 2$ of the ξ_i still spanning V. The process stops when we reach an independent set. Thus:

LEMMA 2-2. If $\xi_1, \ldots, \xi_t$ are nonzero vectors spanning a vector space V over $\mathfrak{F}$, some subset of $\xi_1, \ldots, \xi_t$ forms a basis of V. Moreover, if $\xi_1, \ldots, \xi_s$ $(s \leq t)$ are linearly independent, the subset forming a basis can be chosen to include $\xi_1, \ldots, \xi_s$.

For the final statement of the lemma, notice that the process above deletes only certain vectors which are linear combinations of the preceding ones. No one of $\xi_1, \ldots, \xi_s$ can be deleted by this process.

EXERCISES

1. Prove the independence of the unit vectors of $V_n(\mathfrak{F})$.

★2. If a set of vectors is independent, prove that the set does not include the zero vector.

★3. Prove that a single vector is independent if and only if it is not the zero vector.

★4. If $\xi_1, \ldots, \xi_t$ is a dependent set of vectors, prove that any set containing $\xi_1, \ldots, \xi_t$ is dependent.

5. Determine which of the sets in Exercise 6, Section 2–3, are linearly independent.

6. Show that $\xi_1, \xi_2,$ and ξ_3 in Exercise 3, Section 2–3, form a basis of the subspace V.

7. Find a basis for the subspace V of all vectors (x_1, x_2, x_3, x_4) over $\mathfrak{F}$ satisfying $x_1 + x_2 + x_3 + x_4 = 0$.

8. Let $\xi_1, \ldots, \xi_m$ be a linearly independent set of vectors in $V_n(\mathfrak{F})$. Show that η is a linear combination of the ξ_i if and only if the set $\xi_1, \ldots, \xi_m, \eta,$ is linearly dependent.

9. Prove that the vectors $\xi_1 = (1, 1, 1)$, $\xi_2 = (1, 2, 3)$, and $\xi_3 = (2, 2, 0)$ form a basis of $V_3(\mathfrak{R})$. Find expressions for the unit vectors as linear combinations of these ξ_i.

10. In the following set of vectors in $V_3(\mathfrak{R})$ find the first ξ_m which is a linear combination of the ξ_i preceding it: $\xi_1 = (1, 0, 1)$, $\xi_2 = (0, 2, 3)$, $\xi_3 = (1, 0, 0)$, $\xi_4 = (2, 3, 5)$.

11. Illustrate Lemma 2–2 with the following set of vectors belonging to $V = V_4(\mathfrak{R})$: $\xi_1 = (1, 1, 0, 0)$, $\xi_2 = (1, 0, 1, 0)$, $\xi_3 = (1, 1, 1, 1)$, $\xi_4 = (-1, 0, 0, 1)$.

2–5 Bases of $V_n(\mathfrak{F})$. The unit vectors u_i form a basis for $V_n(\mathfrak{F})$ consisting of n vectors. It will be shown in Theorem 2–4 that every basis of $V_n(\mathfrak{F})$ has n vectors.

THEOREM 2–3. *If $\xi_1, \ldots, \xi_s$ are linearly independent vectors of $V_n(\mathfrak{F})$, there is a basis for $V_n(\mathfrak{F})$ which includes all s of the vectors ξ_i.*

The set of nonzero vectors

$$\text{(11)} \qquad\qquad \xi_1, \ldots, \xi_s, \quad u_1, \ldots, u_n$$

spans $V_n(\mathfrak{F})$. By Lemma 2–2 the set (11) has a subset which includes $\xi_1, \ldots, \xi_s$ and forms a basis of $V_n(\mathfrak{F})$. This proves the theorem. It may be stated briefly as follows: every linearly independent set of vectors in $V_n(\mathfrak{F})$ may be extended to a basis.

THEOREM 2–4. *If a vector space V has a basis, all bases of V include precisely the same number of vectors.*

Proof: Let $\alpha_1, \ldots, \alpha_r$ and $\beta_1, \ldots, \beta_s$ be any two bases of V. Then

$$\text{(12)} \qquad\qquad \alpha_1, \beta_1, \beta_2, \ldots, \beta_s$$

is a dependent set, whence some β_m is a linear combination of the vectors preceding β_m in the list (12).   By a suitable change in notation of the β_j, if necessary, it may be assumed that β_m is β_s.*   Then by Lemma 2–1, V is also spanned by the set

$$(13) \qquad \alpha_1, \beta_1, \ldots, \beta_{s-1}.$$

Since α_2 is thus a linear combination of the vectors in the set (13), the set

$$(14) \qquad \alpha_2, \alpha_1, \beta_1, \ldots, \beta_{s-1}$$

is dependent.   Thus some vector in the set (14) is a linear combination of those preceding it, and this cannot be α_1, since the α_i are independent.   Hence one of the β_i, say β_{s-1} (changing notation again, if necessary), is a combination of the vectors preceding it in the list (14), so that by Lemma 2–1 the set

$$(15) \qquad \alpha_2, \alpha_1, \beta_1, \ldots, \beta_{s-2}$$

spans V.   If $s < r$, a repetition of this process leads to the set

$$(16) \qquad \alpha_s, \alpha_{s-1}, \ldots, \alpha_1$$

as a set spanning V.   Since $s < r$, there is an α_{s+1}, which must be a linear combination of the spanning vectors (16).   But then $\alpha_1, \ldots, \alpha_r$ are dependent, contrary to hypothesis.   Thus s is not less than r, so that $s \geqq r$.   Interchanging the roles of the two bases leads to the like conclusion, $r \geqq s$.   Hence $r = s$.

Since $V_n(\mathfrak{F})$ has a basis consisting of the n unit vectors, Theorem 2–4 implies

COROLLARY 2–4.   *Every basis of $V_n(\mathfrak{F})$ contains precisely n vectors.*

THEOREM 2–5.   *Any set of $n + 1$ vectors of $V_n(\mathfrak{F})$ is dependent.*

Proof: If the set were independent, Theorem 2–3 would provide a basis including these vectors, and hence having more than n vectors. Corollary 2–4 prohibits such bases.

2–6 Dimension of a vector space.   The results of the previous section justify the description of $V_n(\mathfrak{F})$ as a "space of n dimensions," or the definition that the dimension of $V_n(\mathfrak{F})$ is n.   The idea of dimension extends also to subspaces, as we shall now see.

Let V be any nonzero subspace of $V_n(\mathfrak{F})$.   Then V contains a

---

* This is not essential in the argument.   It is merely a convenience in writing down the vectors in the new spanning set (13).

nonzero vector, which is linearly independent. But V cannot contain a set of more than n linearly independent vectors. Thus there must be a maximum integer $r \leqq n$ such that V contains r linearly independent vectors $\xi_1, \ldots, \xi_r$. If ξ is any vector in V, the set

$$\xi_1, \ldots, \xi_r, \xi,$$

containing $r + 1$ vectors, must be dependent, and ξ must be a linear combination of $\xi_1, \ldots, \xi_r$. In other words, $\xi_1, \ldots, \xi_r$ span V, and being linearly independent, form a basis of V. This gives

THEOREM 2–6. *Every nonzero subspace V of $V_n(\mathfrak{F})$ has a basis. All bases of V have the same number of vectors.*

The last statement is implied by Theorem 2–4. The *dimension* of a nonzero space V is defined as the number of vectors in a basis of V. The *dimension* of the zero subspace is defined as zero.

COROLLARY 2–6A. *Every linearly independent set of vectors belonging to V is extendable to a basis of V.*

For proof we need only reread the proof of Theorem 2–3, substituting for $u_1, \ldots, u_n$ a basis $\alpha_1, \ldots, \alpha_r$ of V.

COROLLARY 2–6B. *If V has dimension r, any set of $r + 1$ vectors belonging to V is dependent.*

The proof is parallel to that of Theorem 2–5.

COROLLARY 2–6C. *Let V have dimension r. Then r vectors in V form a basis of V if and only if they are linearly independent.*

If the r vectors are linearly independent but are not a basis, Corollary 2–6A provides a basis with more than r vectors, contrary to Theorem 2–6. Thus the r vectors must form a basis. The converse is trivial.

EXERCISES

1. Show where Theorem 2–5 was used in the proof of Theorem 2–6.

★2. Let r be the dimension of a subspace V of $V_n(\mathfrak{F})$. Show that $V = V_n(\mathfrak{F})$ if and only if $r = n$.

3. In $V_4(\mathfrak{R})$ consider the subspace V of all vectors (x_1, x_2, x_3, x_4) satisfying $x_1 + 2x_2 = x_3 + 2x_4$. Show that the vectors $\xi_1 = (1, 0, 1, 0)$ and $\xi_2 = (0, 1, 0, 1)$ are linearly independent and lie in V. Then extend this set of two vectors to a basis of V.

4. Write a proof of Corollary 2–6A.

5. Write a proof of Corollary 2–6B.

6. Find the dimension of the subspace V of $V_4(\mathfrak{R})$ spanned by $\xi_1 = (1, 1, 2, 3)$, $\xi_2 = (0, 1, 2, 2)$, and $\xi_3 = (3, 2, 4, 7)$.

2-7 $\mathfrak{F}_n$ as a vector space. Although we have written vectors as matrices of one row, we could equally well have written them as columns. In fact all the developments in this chapter are valid if the vectors are written diagonally, sinuously, in T-formation, or in some other shape. The only requirement is that the same shape be employed throughout any discussion*; otherwise addition of vectors would be vague, since we add vectors by adding similarly placed coordinates.

In particular, then, $n \times n$ matrices may be regarded as vectors with m coordinates, where $m = n^2$. If the matrices are written in the usual fashion with n rows and n columns, addition and scalar multiplication may be carried out just as effectively as if the matrices were strung out in long rows:

$$(a_{11}, \ldots, a_{1n}, a_{21}, \ldots, a_{2n}, \ldots, a_{n1}, \ldots, a_{nn}).$$

Thus $\mathfrak{F}_n$, with matrix multiplication neglected, may be regarded as the same thing as $V_m(\mathfrak{F})$, where $m = n^2$ and the vectors of $V_m(\mathfrak{F})$ are written in an unusual manner.

Now let A be an $n \times n$ matrix over $\mathfrak{F}$. Then $I, A, A^2, \ldots, A^m$ are $m + 1$ quantities of $\mathfrak{F}_n = V_m(\mathfrak{F})$. By Corollary 2–6B they must be linearly dependent, so that there are scalars c_i, not all zero, such that

$$c_m A^m + c_{m-1} A^{m-1} + \cdots + c_0 I = 0,$$

where 0 is the zero matrix of $\mathfrak{F}_n$. If c_t is the first nonzero coefficient in this equation, we may multiply by its inverse and obtain an equation of the form,

(17) $$A^t + k_{t-1} A^{t-1} + \cdots + k_1 A + k_0 I = 0.$$

This proves

THEOREM 2–7.   *Each $n \times n$ matrix over $\mathfrak{F}$ satisfies some polynomial equation* (17) *with coefficients k_i in $\mathfrak{F}$.*

---

* Strictly, even this is not necessary. If the n coordinates are written in n colors, for example, we can add vectors by adding "blue" coordinates to get the new "blue" coordinate, and so on.

2–8 Sum of subspaces. In ordinary three-space $V_3(\mathfrak{R})$, consider two distinct lines L_1 and L_2 through the origin. Then there is a unique plane S containing L_1 and L_2. We may say that S is the plane "joining" L_1 and L_2. As usual a vector (x, y, z) may be interpreted either as a point or as a line segment from the origin to the point whose coordinates are x, y, z. Then every point or vector (x, y, z) on S can be shown to be the sum of a pair of vectors, $\xi_1 = (x_1, y_1, z_1)$ on L_1 and $\xi_2 = (x_2, y_2, z_2)$ on L_2, and conversely every such sum lies on S. Since L_1, L_2, and S are subspaces of $V_3(\mathfrak{R})$, these ideas suggest the following important generalizations.

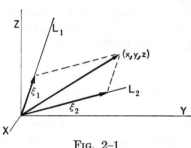

Fig. 2-1

Let V and W be subspaces of $V_n(\mathfrak{F})$, and let $V + W$ denote the set of all vectors $\xi + \eta$, ξ varying over V and η varying over W. The reader may verify that $V + W$ is closed under addition and scalar multiplication. This vector space $V + W$ is called the *sum* or *join* of V and W.

THEOREM 2–8. *If subspaces V and W of $V_n(\mathfrak{F})$ have no vector in common except the zero vector, the dimension of $V + W$ is the sum of the dimensions of V and W.*

Let $\xi_1, \ldots, \xi_v$ and $\eta_1, \ldots, \eta_w$ be bases of V and W, respectively. Taken together these $v + w$ vectors span $V + W$. If a linear combination vanishes, say

$$\Sigma a_i \xi_i + \Sigma b_j \eta_j = 0,$$

then

(18) $$\Sigma a_i \xi_i = -\Sigma b_j \eta_j.$$

The vector represented in two ways in (18) lies in V and in W. By hypothesis this vector must be 0, and

$$\Sigma a_i \xi_i = 0, \quad \Sigma b_j \eta_j = 0,$$

whence every $a_i = 0$ and every $b_j = 0$. This establishes the independence of the $v + w$ vectors ξ_i and η_j, which span $V + W$. Thus they are a basis, and $V + W$ has dimension $v + w$, which is the sum of the dimensions v of V and w of W.

Exercises

1. If n and s are fixed unequal positive integers, let V denote the totality of $n \times s$ matrices over some field $\mathfrak{F}$. Is V a vector space over $\mathfrak{F}$? Can Theorem 2–7 be extended to this case?

2. Prove that $V + W$ is a vector space.

3. Show how Theorem 2–8 is illustrated by L_1, L_2, and S described above.

★4. In Theorem 2–8 show that every vector α in $V + W$ is expressible uniquely in the form $\alpha = \xi + \eta$ with ξ in V and η in W.

5. State and prove the converse of Theorem 2–8.

★6. Let V and W be subspaces of $V_n(\mathfrak{F})$. Show that the set J of all vectors common to V and W is a subspace. This space J is called the *intersection* of V and W, and is denoted by $V \frown W$.

7. Let S and V_1 be subspaces of $V_n(\mathfrak{F})$ such that $V_1 \leqq S$. Show that there is a subspace $V_2 \leqq S$ such that (a) $V_1 + V_2 = S$ and (b) $V_1 \frown V_2 = 0$.

8. Show that properties (a) and (b) in Exercise 7 imply that every vector ξ in S is expressible uniquely in the form $\xi_1 + \xi_2$, ξ_1 in V_1, ξ_2 in V_2. The vector ξ_2 is called the *projection of ξ on V_2 parallel to V_1*, and ξ_1 is the projection of ξ on V_1 parallel to V_2. Apply these ideas to the illustration at the opening of Section 2–8.

CHAPTER 3

EQUIVALENCE, RANK, AND INVERSES

3–1 Equivalent systems of linear equations. A system

$$(1) \qquad \sum_{j=1}^{s} c_{ij}x_j = k_i \qquad (i = 1, \ldots, n)$$

of n linear equations in s unknowns $x_1, \ldots, x_s$, has for a solution a set of s quantities $x_1, \ldots, x_s$ which, on substitution in (1), reduce the left sides to k_i $(i = 1, \ldots, n)$. Thus a solution is a vector belonging to the s-dimensional space $V_s(\mathfrak{F})$. This idea fits in well with the matrix notation for the system. If $C = (c_{ij})$,

$$X = \text{col } (x_1, \ldots, x_s), \quad K = \text{col } (k_1, \ldots, k_n),$$

the system becomes

$$CX = K.$$

In this form the system may be regarded as having only one unknown, the vector X.

Two systems of n linear equations in s unknowns $x_1, \ldots, x_s$ are called *equivalent* if every solution of one system is also a solution of the other, and vice versa. Certain simple operations are commonly used to convert a system to an equivalent one that may be easier to solve. For instance, it surely makes no difference whether an equation is written first or last or in some intermediate position. An equivalent system is, in fact, obtained if

I. Two equations are interchanged.

II. An equation is multiplied by any scalar which has an inverse.

III. An equation, say the ith, is replaced by the sum of the ith equation and k times the jth equation, where $j \neq i$ and k is any scalar.

The justification of operation III, though not so obvious as I and II, is not difficult, and is therefore left as an exercise. The following system, with real number coefficients, will illustrate the use of these operations:

$$\begin{aligned} e_1&: x_1 - 2x_2 + 3x_3 = 6, \\ e_2&: x_1 - x_2 - x_3 = -4, \\ e_3&: 2x_1 + 3x_2 + 5x_3 = 23. \end{aligned}$$

The symbols e_1, e_2, and e_3 will denote the first, second, and third equations, respectively. Further, we shall use symbols like

$$e_2 \to e_2 - e_1, \quad e_3 \to e_3 - 2e_1,$$

to indicate operations of type III replacing the second equation by itself minus the first, and replacing the third equation by itself minus twice the first. These operations give the equivalent system

$$
\begin{aligned}
f_1: x_1 - 2x_2 + 3x_3 &= 6, \\
f_2: x_2 - 4x_3 &= -10, \\
f_3: 7x_2 - x_3 &= 11.
\end{aligned}
$$

Next perform $f_3 \to f_3 - 7f_2$ to produce a new third equation,

$$g_3: 27x_3 = 81.$$

The system now has the form

$$
\begin{aligned}
x_1 - 2x_2 + 3x_3 &= 6, \\
x_2 - 4x_3 &= -10, \\
27x_3 &= 81,
\end{aligned}
$$

from which one readily finds that $x_3 = 3$, and

$$x_2 = -10 + 4x_3 = 2, \quad x_1 = 6 + 2x_2 - 3x_3 = 1.$$

The same operations may be carried out in terms of the 3×4 matrix

$$
A = \begin{bmatrix} 1 & -2 & 3 & 6 \\ 1 & -1 & -1 & -4 \\ 2 & 3 & 5 & 23 \end{bmatrix}.
$$

This is readily recognized as the matrix of coefficients of the original system with the constants attached as a fourth column. It is known as the *augmented matrix* of the given system of equations. The operations performed above on the equations may be performed equally well on the rows r_1, r_2, and r_3 of the matrix $A : r_2 \to r_2 - r_1, r_3 \to r_3 - 2r_1$. The matrix then is

$$
B = \begin{bmatrix} 1 & -2 & 3 & 6 \\ 0 & 1 & -4 & -10 \\ 0 & 7 & -1 & 11 \end{bmatrix}.
$$

Next replace the third row by itself minus seven times the second row, which gives

$$
C = \begin{bmatrix} 1 & -2 & 3 & 6 \\ 0 & 1 & -4 & -10 \\ 0 & 0 & 27 & 81 \end{bmatrix}.
$$

By restoring the unknowns and plus and equal signs in the usual places in this matrix C, we obtain precisely the system from which the solution $(x_1, x_2, x_3) = (1, 2, 3)$ was found. We shall, however, go a bit further.

First multiply the third row of C by $\frac{1}{27}$, then add four times the new third row to the second, and (-3) times the new third row to the first. This produces

$$D = \begin{bmatrix} 1 & -2 & 0 & -3 \\ 0 & 1 & 0 & 2 \\ 0 & 0 & 1 & 3 \end{bmatrix}.$$

Adding twice the second row of D to the first gives

$$M = \begin{bmatrix} 1 & 0 & 0 & 1 \\ 0 & 1 & 0 & 2 \\ 0 & 0 & 1 & 3 \end{bmatrix}.$$

Inserting the unknowns and equal signs, we find at once that $(x_1, x_2, x_3) = (1, 2, 3)$.

3-2 Elementary row operations. With the discussion of Section 3–1 in mind, we define three *elementary row operations* on matrices:

   I. Interchange of two rows.

   II. Multiplication of a row by any scalar* which has an inverse.

   III. Replacement of the ith row by the sum of the ith row and k times the jth row, where $j \neq i$ and k is any scalar.

If $\mathcal{O}_{ij}$ denotes an operation of type I interchanging the ith and jth rows, the original matrix is restored by performing $\mathcal{O}_{ij}$ again. If $\mathcal{O}_i(c)$ denotes an operation of type II, multiplying the ith row by c, $\mathcal{O}_i(c^{-1})$ restores the original matrix. If $\mathcal{O}_{ij}(k)$ denotes an operation of type III, adding k times the jth row to the ith, $\mathcal{O}_{ij}(-k)$ restores the original matrix. In brief, for each elementary row operation $\mathcal{O}$ there is another of the same type which "undoes" the effect of $\mathcal{O}$.

It is profitable to observe that elementary row operations on a matrix A may be accomplished by forming matrix products EA with suitable matrices E. To accomplish an operation $\mathcal{O}$ on an $n \times s$ matrix A, use for E the matrix obtained by performing $\mathcal{O}$ on the matrix I_n. In the example given in Section 3–1, if we wish to inter-

---

* That is, any nonzero scalar. The wording of II has been chosen with a view towards a later situation.

change the first and second rows, we do this to I_3, finding a matrix E_{12}, then calculate $E_{12}A$:

$$E_{12} = \begin{bmatrix} 0 & 1 & 0 \\ 1 & 0 & 0 \\ 0 & 0 & 1 \end{bmatrix}, \quad E_{12}A = \begin{bmatrix} 1 & -1 & -1 & -4 \\ 1 & -2 & 3 & 6 \\ 2 & 3 & 5 & 23 \end{bmatrix}.$$

To add -7 times the second row of B (in Section 3–1) to the third row, do this to I_3, obtaining a matrix

$$E_{32}(-7) = \begin{bmatrix} 1 & 0 & 0 \\ 0 & 1 & 0 \\ 0 & -7 & 1 \end{bmatrix};$$

then calculate

$$E_{32}(-7)B = \begin{bmatrix} 1 & -2 & 3 & 6 \\ 0 & 1 & -4 & -10 \\ 0 & 0 & 27 & 81 \end{bmatrix} = C.$$

To multiply row 3 in C by $\frac{1}{27}$, do this to I_3 first:

$$E_3\left(\tfrac{1}{27}\right) = \begin{bmatrix} 1 & 0 & 0 \\ 0 & 1 & 0 \\ 0 & 0 & \frac{1}{27} \end{bmatrix};$$

then calculate

$$E_3\left(\tfrac{1}{27}\right)C = \begin{bmatrix} 1 & -2 & 3 & 6 \\ 0 & 1 & -4 & -10 \\ 0 & 0 & 1 & 3 \end{bmatrix}.$$

LEMMA 3–1. *To perform an elementary row operation $\odot$ on an $n \times s$ matrix A, calculate the product EA, where E is the matrix obtained by performing $\odot$ on I_n.*

To prove the lemma, let A be partitioned into its rows $A_1, \ldots, A_n$, so that each row of EA is a linear combination of these rows as asserted in Theorem 1–5. Notice that row h of I is the unit vector u_h, and that

$$u_h A = A_h.$$

If $\odot$ is the operation $\odot_{ij}$ interchanging rows i and j, in the matrix E we find that

$$\text{row } i = u_j, \quad \text{row } j = u_i.$$

Then in EA

$$\text{row } i = u_j A = A_j, \quad \text{row } j = u_i A = A_i$$

as desired, and if $h \neq i, j$, row h of EA is $u_h A = A_h$. This proves the lemma for operations $\mathcal{O} = \mathcal{O}_{ij}$. If $\mathcal{O} = \mathcal{O}_{ij}(k)$, then in E

$$\text{row } i = u_i + k u_j, \quad \text{row } h = u_h. \qquad (h \neq i)$$

It follows that row h of EA is A_h if $h \neq i$, and row i of EA is

$$(u_i + k u_j)A = u_i A + k \cdot u_j A$$
$$= A_i + k \cdot A_j,$$

as desired. If $\mathcal{O} = \mathcal{O}_i(c)$ we must multiply row i of A by c. Then E is the same as I except for the element c as the ith diagonal element; and it is clear that EA is the same as matrix A, except for the fact that cA_i is the ith row of EA.

Any matrix E obtained by performing a single elementary row operation on I is called an *elementary matrix*. For the operations $\mathcal{O}_{ij}$, $\mathcal{O}_i(c)$, and $\mathcal{O}_{ij}(k)$, let the corresponding elementary matrices be denoted by E_{ij}, $E_i(c)$, and $E_{ij}(k)$. Then

$$(2) \qquad\qquad E_{ij} E_{ij} = I,$$

that is, if starting with I we interchange ith and jth rows, then on the result perform the same interchange, we get back to I. Similarly,

$$(3) \qquad\qquad E_i(c^{-1}) E_i(c) = I = E_i(c) E_i(c^{-1}),$$
$$(4) \qquad\qquad E_{ij}(-k) E_{ij}(k) = I = E_{ij}(k) E_{ij}(-k).$$

These equations (2), (3), and (4) may be interpreted thus: Given an elementary matrix E, there exists a matrix D such that $DE = I = ED$.

DEFINITION 1. A matrix A is said to be *nonsingular* if there is a matrix B such that

$$BA = I = AB.$$

Any such matrix B is called an *inverse* of A. If there is no such matrix B, then A is called *singular*.

To say that A has an inverse is to say that A is nonsingular. Many matrices do not have inverses. However, the remarks above the definition prove

THEOREM 3–1. *Every elementary matrix has an inverse, which is also elementary.*

The size of a nonsingular matrix A is not arbitrary. Suppose that A is $n \times s$, so that the equation $AB = I$ implies that $I = I_n$. Then B must be $s \times n$. But also $BA = I_n$, and BA is $s \times s$, so that $s = n$. This proves the first part of the next theorem.

THEOREM 3-2. *If A is nonsingular it must be square, and its inverse is unique.*

To prove the last part, let C and B be inverses of A. Then

$$C(AB) = CI = C = (CA)B = IB = B,$$

so that

$$C = B.$$

The inverse of a matrix A is denoted by A^{-1}.

Attention is called to the following useful properties, to be verified in the exercises. If A and B have inverses, so do AB and A', and

(5) $$(AB)^{-1} = B^{-1}A^{-1}, \quad (A')^{-1} = (A^{-1})'.$$

Stated differently, a product of nonsingular matrices is nonsingular, and the inverse of the product is the product of the inverses in reverse order. Also, the transpose of a nonsingular matrix is nonsingular, and the inverse of the transpose is the transpose of the inverse.

EXERCISES

1. If

$$A = \begin{bmatrix} 1 & 2 & 3 \\ 4 & 5 & 6 \\ 7 & 8 & 9 \end{bmatrix},$$

use matrix multiplication to calculate $\mathcal{O}_{12}A$, $\mathcal{O}_{12}(-2)A$, $\mathcal{O}_3(2)A$, where $\mathcal{O}A$ denotes the result of performing operation $\mathcal{O}$ on A.

2. Write the inverses of the 3×3 elementary matrices E_{12}, $E_{23}(5)$, $E_3(2)$, and verify each inverse E^{-1} by calculating the product $EE^{-1} = I_3$.

★3. Prove the formulas (5) for the inverse of a product and the inverse of the transpose of a matrix.

★4. If A_i is an $n \times n$ nonsingular matrix $(i = 1, \ldots, r)$, state and prove a formula for $(A_1 \cdots A_r)^{-1}$ in terms of the matrices A_i^{-1}.

5. Let A be a square matrix. Prove that A does not have an inverse if any row of A is a zero vector, or if two rows of A are identical.

★6. For each positive integer r define $A^{-r} = (A^{-1})^r$, A being a nonsingular matrix. Also define $A^0 = I$. Then prove the laws of exponents: $A^r A^s = A^{r+s}$, $(A^r)^s = A^{rs}$ for exponents r and s which are arbitrary integers.

★7. If A commutes with B, and A is nonsingular, prove that A^{-1} commutes with B.

★8. Let V be a subspace of $V_n(\mathfrak{F})$, whose vectors are to be written as columns. Let A be an $n \times n$ matrix over $\mathfrak{F}$. Prove that the vectors $A\xi$, with ξ varying over V, form a subspace of $V_n(\mathfrak{F})$, and that this subspace has the same dimension as V if A is nonsingular.

3-3 Row equivalence. A matrix B obtained by performing a succession of elementary row operations on a rectangular matrix A is said to be *row equivalent* to A. When it is necessary to make clear that $\mathfrak{F}$ is the field of scalars for the matrices concerned and that the scalars for the row operations of type II and type III are to be chosen in $\mathfrak{F}$, we say that B is row equivalent *over $\mathfrak{F}$* to A.

The investigation of how far a matrix may be simplified by elementary row operations may be reduced to the following question: What is the simplest matrix in the class of all matrices row equivalent to A? The answer varies, since the word "simplest" is subject to personal interpretation. One common answer is given now. The phrases *zero row* and *zero column* mean that every element of the row or column is zero. A nonzero row, then, is a row with at least one nonzero element.

THEOREM 3-3. *Every rectangular matrix A is row equivalent to a matrix B in which*

(a) *the first r rows, for some $r \geq 0$, are nonzero and all remaining rows, if any, are zero;*

(b) *in the ith row ($i = 1, 2, \ldots, r$) the first nonzero element is equal to unity, the column in which it occurs being numbered c_i;*

(c) $c_1 < c_2 < \cdots < c_r$;

(d) *in column c_i the only nonzero element is the 1 in row i.*

An example of a matrix with these four properties is

$$B = \begin{bmatrix} 0 & 1 & 3 & 5 & 0 & 0 & 1 & 3 \\ 0 & 0 & 0 & 0 & 1 & 0 & 4 & 2 \\ 0 & 0 & 0 & 0 & 0 & 1 & 3 & 7 \\ 0 & 0 & 0 & 0 & 0 & 0 & 0 & 0 \\ 0 & 0 & 0 & 0 & 0 & 0 & 0 & 0 \end{bmatrix}.$$

To prove the theorem, let c_1 be the number of the first nonzero column of A. (If there is no such column, $A = 0$ is already the desired matrix B.) In column c_1 a nonzero element occurs in some row, and if this is not the first row it may be interchanged with the first row (an operation of type I). This nonzero element, which now lies in row 1 and column c_1, may be converted to unity by an operation of type II. If any other row, say the jth, has a nonzero element in column c_1, this element k can be changed to zero by an operation $\mathcal{O}_{j1}(-k)$. In this way all of column c_1 can be made zero, except for the 1 in row 1.

If in the matrix A_1 thus obtained the rows below the first are all zero, the proof is complete, and A_1 is the desired matrix B. Otherwise, choose the first column of A_1 which has a nonzero element c below row 1, and let c_2 denote the number of this column. By an interchange of rows, column c_2 can be made to have this nonzero element c in row 2, and c can be converted to 1 by multiplying row 2 by c^{-1}. All other elements in column c_2 can now be made 0 by adding suitable multiples of the second row to the other rows. Throughout these operations, column c_1 does not change its appearance at all.

A repetition of this process eventually leads to a stage in which column c_r, for some r, has been made all zero except for an element unity in row r, and either row r is the bottom row or every row below it is zero. The matrix then at hand is the matrix B of Theorem 3-3.

COROLLARY 3-3. *Let A and B be the matrices of Theorem 3-3, and let A_1 and B_1 be the submatrices of A and B, respectively, obtained by deleting the last columns of these matrices. Then B_1 is row equivalent to A_1, and B_1 has the properties (a)–(d).*

That B_1 is row equivalent to A_1 is evident from the definition of row operations. It may have a smaller number of nonzero rows than has B. This would be true, for example, if B were the identity matrix. Each of the properties (a) through (d) is clearly true for B_1.

EXERCISES

★1. Let A, B, and C be $n \times s$ matrices over $\mathfrak{F}$. Prove that
    (a) A is row equivalent to A.
    (b) If A is row equivalent to B, B is row equivalent to A.
    (c) If C is row equivalent to B, and B to A, then C is row equivalent to A.

2. For each of the following matrices A over $\mathfrak{R}$ find the corresponding matrix B of Theorem 3-3:

$$\begin{bmatrix} 1 & 2 & 3 \\ 4 & 5 & 6 \\ 7 & 8 & 9 \end{bmatrix}, \quad \begin{bmatrix} 1 & 0 & 1 & 4 \\ 2 & 0 & 1 & 2 \\ -3 & 1 & 2 & 1 \end{bmatrix}, \quad \begin{bmatrix} 1 & 2 & 3 \\ 0 & 1 & 4 \\ 2 & 3 & 1 \\ 3 & 1 & 2 \end{bmatrix}.$$

★3. Prove that the matrix B of Theorem 3-3 is uniquely determined by the given matrix A.

3-4 Row space. If A is an $n \times s$ matrix over $\mathfrak{F}$, its rows are vectors belonging to $V_s(\mathfrak{F})$. All n rows of A then span a subspace V

called the *row space* of A; the dimension of V is called the *row rank* of A. In this and the following sections we begin to reap the profits from Chapter 2.

Notice first that each linear combination of the rows of A with coefficients $g_1, \ldots, g_n$ is a matrix product GA, where $G = (g_1, \ldots, g_n)$, and the row space V of A is the totality of such products GA as G varies over $V_n(\mathfrak{F})$. Since (Theorem 1–5) each row of a product PA is one of these products GA, the rows of PA lie in V. Then linear combinations of the rows of PA are simply linear combinations of vectors belonging to V, so that

(6)   row space of PA $\leq$ row space of A.

This is the first part of the following result.

Lemma 3–2. *The row space of a matrix product PA is contained in that of A. If P is nonsingular, the row space of PA coincides with that of A.*

In the last part of the lemma, P has an inverse P^{-1}. If the result in (6) is applied with PA as the given matrix and P^{-1} as the left multiplier,

row space A = row space $P^{-1}(PA)$ $\leq$ row space PA.

When combined with (6) this inclusion proves that A and PA have the same row space.

Since elementary row operations are accomplished by left multiplication by matrices having inverses, Lemma 3–2 has the following immediate consequence.

Theorem 3–4. *Elementary row operations do not change the row space of a matrix.*

A matrix B in the simplified form of Theorem 3–3 has for its row rank the number r of nonzero rows. For, suppose these row vectors are designated in order by $\xi_1, \ldots, \xi_r$, and suppose that $k_1\xi_1 + \cdots + k_r\xi_r = 0$. The c_1th coordinate of this vector must be k_1 by the fact that ξ_1 has a 1 in column c_1 and all the other ξ_i have zeros there. Hence $k_1 = 0$, and in similar fashion every $k_i = 0$. Thus the row space of B is spanned by r vectors which are linearly independent, so that these vectors form a basis, and r is the row rank of B. By the last theorem this is also the row rank of any matrix A row equivalent to B. Thus

THEOREM 3-5. *The row rank of a matrix A over $\mathfrak{F}$ is the number of nonzero rows in the row equivalent matrix B of Theorem 3-3.*

We have not proved that the matrix B of Theorem 3-3 is uniquely determined by the given matrix A, but Theorem 3-5 shows that the number of nonzero rows in B is unique; it is the row rank of A. In fact, B itself is unique.

EXERCISES

1. Is the converse of the second statement in Lemma 3-2 true? Hint: Use 2×2 matrices over $\mathfrak{R}$ for examples.

2. Let a subspace V of $V_n(\mathfrak{F})$ have dimension r. Let $\alpha_1, \ldots, \alpha_s$ belong to V, and let M be the $s \times n$ matrix having α_i as its ith row, $i = 1, \ldots, s$. Show that $\alpha_1, \ldots, \alpha_s$ span V if and only if M has row rank r.

3. Prove that the following matrices have the same row space V, then find a basis for V:

$$\begin{bmatrix} 2 & 3 & 4 \\ 0 & 1 & 1 \\ 2 & -1 & 0 \end{bmatrix}, \quad \begin{bmatrix} 4 & 2 & 4 \\ 2 & 0 & 1 \\ 2 & 7 & 8 \end{bmatrix}.$$

3-5 Solution of linear systems. Since elementary row operations are precisely the operations employed in solving systems of simultaneous linear equations, Theorem 3-3 may be applied to such systems. Consider the system $CX = K$, that is,

$$(7) \qquad \sum_{j=1}^{s} c_{ij}x_j = k_i. \qquad (i = 1, \ldots, n)$$

Here $C = (c_{ij})$ is the $n \times s$ coefficient matrix of the system, $X = \text{col } (x_1, \ldots, x_s)$ is the column of unknowns, and $K = \text{col } (k_1, \ldots, k_n)$ is the column of constants. Let A denote the $n \times (s + 1)$ *augmented matrix*

$$A = (C, K)$$

of the system, obtained from C by attaching the column of constants as an extra column. We shall prove

THEOREM 3-6. *A system of simultaneous linear equations has a solution if and only if the rank of the augmented matrix equals the rank of the coefficient matrix.[*]*

---

[*] Here and elsewhere the term *row rank* is shortened to *rank*.

As we have seen, the system may be reduced to one whose augmented matrix has the simplified form described by properties (a)–(d) of Theorem 3–3, the new and old augmented matrices having the same rank r. Corollary 3–3 then assures us that the new and old coefficient matrices are row equivalent, hence have the same rank, c, and that the new coefficient matrix also has the properties (a)–(d) of Theorem 3–3. In proving the theorem above we may assume at the outset that the matrices are in the simplified form of Theorem 3–3. Then (Theorem 3–5) the row ranks of C and $A = (C, K)$ are their numbers of nonzero rows. Since each nonzero row of C is part of a nonzero row of A, we have

$$r = \text{row rank of } (C, K) \geq c = \text{row rank of } C.$$

Suppose, first, that $r > c$. Then row $c + 1$ in C is zero, but row $c + 1$ in A is not zero. Thus equation $c + 1$ has the form

$$0x_1 + \cdots + 0x_s = k_{c+1} \neq 0.$$

Since this equation cannot be satisfied by any choice of the unknowns, the system is *inconsistent*, that is, has no solution.

We now take up the only remaining possibility, namely that $r = c$, and will find in this case that the difficulty above does not occur. By properties (a)–(d) of Theorem 3–3 there are r unknowns

$$x_{c_1}, \ldots, x_{c_r},$$

such that x_{c_i} occurs with nonzero coefficient only in equation i, this nonzero coefficient being unity. An example of such a system is the following:

$$
\begin{aligned}
x_1 + 5x_2 \quad\;\; + 6x_4 \quad\;\; &= 2, \\
x_3 + \;\; x_4 \quad\;\; &= 1, \\
x_5 &= 3.
\end{aligned}
$$

Here we have $r = 3 = c$ and

$$c_1 = 1, \quad c_2 = 3, \quad c_3 = 5.$$

If we bring to the right the terms containing those x_j with j different from the numbers 1, 3, 5 above, the system becomes

(8)
$$
\begin{aligned}
x_1 &= 2 - 5x_2 - 6x_4, \\
x_3 &= 1 - x_4, \\
x_5 &= 3.
\end{aligned}
$$

Thus we have solved for x_1, x_3, and x_5 in terms of the remaining unknowns. which may be called parameters; all solutions $(x_1, x_2, x_3,$

x_4, x_5) are obtained by assigning arbitrary values to the parameters x_2 and x_4, then determining values of x_1, x_3, and x_5 from equations (8). The system is surely consistent, that is, has a solution.

These ideas clearly carry over to the general case. Let the terms containing those unknowns x_j with $j \neq c_1$, ..., c_r be transferred to the right, so that the left side of equation i is merely the term x_{c_i} ($i = 1$, ..., r). We have then solved for

$$x_{c_1}, \ldots, x_{c_r}$$

in terms of the remaining unknowns, which may be regarded as parameters. The system is consistent, and all solutions are found by assigning arbitrary values to the parameters and determining the values of x_{c_1}, ..., x_{c_r}. This completes the proof of the theorem.

A system (7) is called *homogeneous* if all of the constant terms k_i are zero. The consistency of a homogeneous system is guaranteed not only by the fact that its augmented and coefficient matrices have the same rank, but also by the fact that it has the solution $(x_1, \ldots, x_s)$ $= (0, \ldots, 0)$ in which every x_i is zero. This solution, the zero vector, is called *trivial*. Thus the question of importance for homogeneous systems concerns the existence of a nontrivial solution. By continuing the analysis above we may prove the following criterion.

COROLLARY 3–6. *A homogeneous system of linear equations has a nontrivial solution if and only if the number of unknowns exceeds the row rank of the coefficient matrix.*

As above, let r denote the common rank of the coefficient and augmented matrices, and let s denote the number of unknowns. There are r unknowns

$$x_{c_1}, \ldots, x_{c_r}$$

for which we may solve in terms of the remaining unknowns or parameters x_j, if any. There will actually be some of these parameters x_j provided that $s > r$. Since arbitrary values for these parameters x_j form part of a solution, we take these x_j equal to 1 and have a nontrivial solution. Thus the condition $s > r$ implies the existence of a nontrivial solution.

The only remaining possibility is that the r special unknowns x_{c_1}, ..., x_{c_r} shall be the full number of unknowns, that is, that $s = r$. In this case no unknowns remain to be transferred to the right side. Moreover, by property (c) of Theorem 3–3,

$$1 \leqq c_1 < c_2 < \cdots < c_r \leqq r.$$

It follows that

$$c_1 = 1, \quad c_2 = 2, \ldots, \quad c_r = r.$$

The ith equation begins with the term $x_{c_i} = x_i$, and contains no term $x_{c_j}, j \neq i$, by property (d) of Theorem 3–3. Thus equation i contains only the term x_i and must be the equation

$$x_i = 0. \qquad\qquad (i = 1, \ldots, r)$$

The assumption that $r = s$ thus leads to the conclusion that only the trivial solution is possible. This covers all the possibilities and completes the proof.

EXERCISES

1. Solve each of the following systems by reducing the augmented matrix of each system to the simplified form of Theorem 3–3.

(a) $\begin{cases} x_1 - x_2 + x_3 = 3, \\ 2x_1 + x_2 + 2x_3 = 3, \\ 3x_1 + 2x_2 - x_3 = 0. \end{cases}$

(b) $\begin{cases} x_1 - x_2 + x_3 = 3, \\ 2x_1 + x_2 + 2x_3 = 3, \\ x_1 + 2x_2 - x_3 = 0. \end{cases}$

(c) $\begin{cases} 2x_1 + x_2 + x_3 = 8, \\ x_1 + x_2 + 3x_3 = 10. \end{cases}$

(d) $\begin{cases} 2x_1 + x_2 + x_3 = 0, \\ x_1 + x_2 + 3x_3 = 0. \end{cases}$

2. Without finding any solutions, show how to prove that the system (d) in Exercise 1 must have a nontrivial solution.

3. Without solving, determine whether the following system has a nontrivial solution:

$$x_1 + x_2 + x_3 = 0,$$
$$x_1 - x_2 + x_3 = 0,$$
$$x_1 - x_2 - x_3 = 0.$$

★4. Let the vector X_0 be a fixed solution of the system (7). Show that all solutions of the system are given by $X_0 + X_h$ as X_h varies over all solutions X of the corresponding homogeneous system, $CX = 0$.

5. Without attempting to solve, prove that the following system over $\Re$ is inconsistent:

$$x_1 - x_2 + x_3 = 2,$$
$$x_1 + x_2 - x_3 = 1,$$
$$x_1 + 5x_2 - 5x_3 = 1.$$

Then use the solution process of Exercise 1 to see how the inconsistency arises in the last simplified equation.

6. Suppose that the coefficient matrix of a system of n linear equations in s unknowns has row rank n. Prove that the system is consistent.

7. Suppose that a system $CX = K$ of n linear equations in n unknowns has a unique solution, that is, that there is one and only one vector X_0 satisfying the condition $CX_0 = K$. Prove that the system remains consistent no matter how the vector K of constants is altered, and that each system obtained has only one solution.

3-6 Column operations. The columns of an $n \times s$ matrix A over $\mathfrak{F}$ are vectors in $V_n(\mathfrak{F})$, and the subspace which they generate is the *column space* of A. Its dimension is called the *column rank* of A. Elementary *column operations* may also be defined as the interchange of two columns, the multiplication of a column by a scalar which has an inverse, and the addition to one column of any scalar multiple of a different column. The symbols

$$(9) \qquad \mathcal{CO}_{ij}, \quad \mathcal{CO}_i(c), \quad \mathcal{CO}_{ij}(k)$$

will denote these operations. The first of the symbols (9) means that column i and column j are to be interchanged. The second symbol indicates that column i is to be multiplied by c. The last symbol in (9) indicates that we are to add to column i the product k times column j. Each of the operations (9) determines a unique corresponding row operation — simply omit the $\mathcal{C}$:

$$\mathcal{CO}_{ij} \quad \leftrightarrow \mathcal{O}_{ij},$$
$$\mathcal{CO}_i(c) \quad \leftrightarrow \mathcal{O}_i(c),$$
$$\mathcal{CO}_{ij}(k) \leftrightarrow \mathcal{O}_{ij}(k).$$

All of the foregoing theory of row operations (Lemmas 3–1, 3–2, Theorems 3–3, 3–4, and 3–5) can be paralleled perfectly with column operations. Some of these parallels are listed below.

LEMMA 3–3. (i) *Each elementary column operation $\mathcal{CO}$ on arbitrary $n \times s$ matrices A can be achieved by multiplying A on the right by the matrix E which is obtained by performing $\mathcal{CO}$ on I_s. Moreover, E is the transpose of the matrix which effects the corresponding row operation $\mathcal{O}$ by multiplication on the left.* (ii) *If Q is nonsingular, the column spaces of A and AQ coincide.* (iii) *Elementary column operations do not change the column space of a matrix.*

Although these results, and also the column analogues of Theorems 3–3 and 3–5, can be proved in the same ways as the corresponding results for rows, a simpler device is available. The process of transposing a matrix can be used to prove these "column" results as consequences of the corresponding row results. To prove (i) note that the columns of A appear in A' as rows. If we perform $\odot$ on A', then transpose, the resulting matrix $(\odot A')'$ is the desired matrix $\mathfrak{C}\odot A$. But operation $\odot$ can be achieved by multiplying on the left by an elementary matrix E_0, so that

$$\mathfrak{C}\odot A = (E_0 A')' = A E_0'.$$

This shows that operation $\mathfrak{C}\odot$ can be achieved by multiplication on the right by the matrix $E = E_0'$. Taking A to be the identity matrix now gives

$$\mathfrak{C}\odot I = IE = E,$$

so that E is in fact obtainable from I by performing operation $\mathfrak{C}\odot$. This proves (i).

To prove (ii) note that the column spaces of A and AQ are, respectively, the same as the row spaces of A' and $(AQ)' = Q'A'$. Since Q' is nonsingular, these spaces coincide by Lemma 3–2, completing the proof. Property (iii) is an immediate consequence of (i) and (ii).

We have defined elementary matrices as the matrices obtained by performing single elementary row operations on an identity matrix. If column operations had been specified instead of row operations, the class of matrices so constructed would coincide with the class of elementary matrices as originally defined. To prove this assertion we refer to the second sentence of (i) in the last lemma, which informs us that the new class consists of the transposes of the matrices in the original class. However, each elementary matrix E_{ij} or $E_i(c)$ is equal to its own transpose. To illustrate the situation for $E_{ij}(k)$ take the following 3×3 case:

$$E_{13}(k) = \begin{bmatrix} 1 & 0 & k \\ 0 & 1 & 0 \\ 0 & 0 & 1 \end{bmatrix} = E_{31}(k)'.$$

In general, $E_{ij}(k)$ is an identity matrix modified to have the element k in the (i, j) place. Hence its transpose has this element k in the (j, i) place and thus equals $E_{ji}(k)$. This proves the assertion above that elementary matrices are definable in terms of row or column operations.

Danger! Do not conclude that each elementary column operation on an arbitrary matrix A may be achieved by performing suitable row operations. Prove that the second of the matrices

$$\begin{bmatrix} 1 & 0 \\ 0 & 0 \end{bmatrix}, \quad \begin{bmatrix} 0 & 1 \\ 0 & 0 \end{bmatrix},$$

is obtainable from the first by an elementary column operation, but not by any succession of elementary row operations whatsoever.

EXERCISES

1. Do the problem stated immediately above.
2. Prove (iii) of Lemma 3–3 by the method illustrated, avoiding use of (ii).
3. State carefully the column analogues of Theorems 3–3 and 3–5.
4. Let $A = (a_{ij})$ be a general 3×3 matrix. By multiplying by a suitable elementary matrix E add k times the third column of A to its first column. Then show by direct multiplication that E' effects the corresponding row operation.

3-7 Equivalence. Elementary row and column operations will be referred to simply as *elementary operations*. Any matrix B obtained from A by performing a succession of elementary operations is said to be *equivalent* to A. Then B is equivalent to A if and only if $B = PAQ$, where P and Q are products* of elementary matrices.

Evidently (1) each matrix A is equivalent to itself, since $\mathcal{O}_1(1)$ performed on A produces A. Moreover, (2) if B is equivalent to A, A is also equivalent to B; and (3) if C is equivalent to B, and B to A, then C is equivalent to A. These properties are useful in many ways. For one thing, by (2) we no longer have to distinguish between the assertions that B is equivalent to A and that A is equivalent to B. Also, if A and C are both equivalent to B, they are equivalent to each other, as one can prove easily. Thus the phrase "is equivalent to" has very much the force of "is equal to."

Let us look at this matter once more. Consider the class of all $n \times s$ matrices over $\mathfrak{F}$, n and s being fixed. Equivalence is a relation between ordered pairs of matrices in this class: given A and B, either B is in the relation of equivalence to A or it is not; although actual testing in numerical cases may be difficult, there can be no doubt that careful investigators will get the same answer. This relation of

---

* It is understood that the "products" may have only one factor. Moreover, P or Q or both may be I.

equivalence, to repeat what was said above, obeys the following three laws:

REFLEXIVE LAW: Each matrix A is in the relation to itself.

SYMMETRIC LAW: If B is in the relation to A, A must be in the relation to B.

TRANSITIVE LAW: If C is in the relation to B, and B to A, then C must be in the relation to A.

DEFINITION 2. Suppose a relation is defined on a class of matrices in such a way that for each ordered pair of matrices in the class the relation either holds or fails. Then if the relation obeys the reflexive, symmetric, and transitive laws, it is called an *RST relation*.*

Equivalence is but one of many RST relations to be studied in this book. The first major result on equivalence is almost at hand. It is begun in Theorem 3–7 and completed in Theorem 3–11.

THEOREM 3–7. *Each matrix A is equivalent to a matrix B, in which for some integer $r \geqq 0$ the first r elements on the diagonal are 1, and all other elements of the matrix are 0. Both the row rank and the column rank of B are equal to r.*

For proof, start with the simplified matrix PA of Theorem 3–3. The columns $c_1, \ldots, c_r$ of PA, all of which are unit vectors, can by column interchanges be brought into the first r positions. This gives a matrix of the form

$$\begin{bmatrix} I_r & D \\ 0 & 0 \end{bmatrix},$$

where the 0's denote zero matrices of appropriate sizes, or are entirely absent if r is the full number of rows. The matrix D can be converted to 0 by adding to its columns suitable multiples of the first r columns. This gives the matrix B of Theorem 3–7. It is clear that B has column rank = row rank = r.

If two attempts to reduce A to matrices fulfilling the description of B in Theorem 3–7 led to "answers" B_1 and B_2, they could differ at most in their numbers, r_1 and r_2, of ones on the diagonal. That $r_1 = r_2$, so that the integer r of Theorem 3–7 is uniquely determined by the given matrix A, will be proved after the concepts in the next section have been developed.

---

* This is commonly called an "equivalence relation," but in the present context such terminology would be confusing.

EXERCISES

★1. Prove the reflexive, symmetric, and transitive laws for equivalence.

2. For each of the following matrices A find the corresponding matrix B of Theorem 3-7:

$$\begin{bmatrix} 1 & 2 & 3 \\ 3 & 2 & 1 \\ 1 & 1 & 1 \end{bmatrix}, \quad \begin{bmatrix} 2 & 4 & 6 & 8 \\ 1 & 0 & 2 & 0 \\ 3 & 1 & 2 & 1 \end{bmatrix}.$$

In each case find matrices P and Q such that $PAQ = B$, where P and Q are products of elementary matrices.

3. Let A be a matrix on which we wish to perform one row operation and one column operation. Show that the same result is obtained regardless of which operation is performed first. Hint: Perform the operations by matrix multiplication.

4. If A and C are equivalent to a common matrix, prove that they are equivalent to each other. Can this property of equivalence be generalized to any RST relation?

3-8 Null space. If A is an $n \times s$ matrix over $\mathfrak{F}$, there may be an $s \times 1$ matrix $X = \text{col}(x_1, \ldots, x_s)$ over $\mathfrak{F}$ such that

(10) $$AX = 0.$$

For example,

$$A = \begin{bmatrix} 1 & 2 & 1 \\ 2 & 1 & 1 \end{bmatrix}, \quad X = \text{col}(1, 1, -3), \quad AX = \begin{bmatrix} 0 \\ 0 \end{bmatrix}.$$

In computing AX one should think of A as partitioned into its columns $A^{(1)}, \ldots, A^{(s)}$, whence (as in Theorem 1-5)

$$AX = x_1 A^{(1)} + \cdots + x_s A^{(s)}.$$

The vectors X belong to $V_s(\mathfrak{F})$, regarded throughout this section as a set of columns, and the totality N of such vectors X satisfying (10) will be shown to be a subspace of $V_s(\mathfrak{F})$. First, N contains at least one vector, the zero vector. If X and Y belong to N,

$$A(X + Y) = AX + AY = 0,$$

and N is closed under addition. Last, if k is any scalar and X lies in N,

$$A(kX) = k(AX) = 0,$$

so that kX lies in N, and N is a subspace. This subspace is called the *null space* of A and its dimension is the *nullity* of A. When convenient we shall write $N(A)$ for the null space of A.

THEOREM 3–8. *Let A be the coefficient matrix of a system of n homogeneous, linear equations in s unknowns. If A has nullity q, the system has q linearly independent solutions $X_1, \ldots, X_q$ such that every solution is a linear combination $a_1X_1 + \cdots + a_qX_q$, and every such combination is a solution.*

The equation (10) is a system of homogeneous, linear equations in $x_1, \ldots, x_s$, and N is simply the totality of solutions of the system. The theorem follows from taking $X_1, \ldots, X_q$ as a basis for N.

THEOREM 3–9. *For any matrix A, column rank + nullity = number of columns.*

The 2×3 matrix A in the illustration above has column rank 2 and number of columns equal to 3. Thus its nullity must be 1. In general let A be $n \times s$ and let its nullity be q. Then its null space N is a subspace of $V_s(\mathfrak{F})$ and has a basis $\nu_1, \ldots, \nu_q$. Since this set is linearly independent, it may (Theorem 2–3) be extended to a set

$$(11) \qquad \nu_1, \ldots, \nu_q, \quad \rho_1, \ldots, \rho_r,$$

which is a basis of $V_s(\mathfrak{F})$. Then

$$(12) \qquad r + q = s,$$

and it remains only to show that r is the column rank of A. Every vector X in $V_s(\mathfrak{F})$ may be written in the form

$$X = \sum_{i=1}^{q} a_i\nu_i + \sum_{j=1}^{r} b_j\rho_j,$$

whence, since each product $A\nu_i = 0$,

$$AX = \Sigma Ab_j\rho_j = b_1(A\rho_1) + \cdots + b_r(A\rho_r).$$

Since AX is an arbitrary vector in the column space of A, this equation shows that the r vectors $A\rho_j$ span this column space. These vectors also are linearly independent. For, if

$$k_1(A\rho_1) + \cdots + k_r(A\rho_r) = 0,$$

then

$$A\xi = 0, \quad \xi = k_1\rho_1 + \cdots + k_r\rho_r,$$

so ξ belongs to N. Then ξ is a linear combination

$$\xi = \Sigma h_i\nu_i = \Sigma k_j\rho_j.$$

This property violates the unique expressibility of vectors in terms of the set (11) unless all the k_j and h_i are 0. This completes the proof that $A\rho_1, \ldots, A\rho_r$ are linearly independent and thus a basis of the column space of A. Then r is the column rank of A, and (12) gives the theorem.

This result has geometrical overtones by virtue of the fact that the solutions of a system (10) may be regarded as the locus (in space of s dimensions) of the system. Suppose, for example, that $s = 3$ and $n = 2$ as in the matrix A below (10). Then the locus of the system, $AX = 0$, of two equations is a geometrical configuration of dimension $q = 3 - r$. Since the coefficient matrix A has column rank $r = 2$, the locus is one-dimensional, a line. The student can verify this assertion independently of matrix theory by showing that the two equations represent nonparallel planes, so that their intersection is a line. If r were 1, the two planes would coincide, and the locus would be a plane in keeping with the dimension formula $q = 3 - r = 2$.

LEMMA 3–4. *If P is nonsingular, A and PA have the same column rank.*

Since A and PA have the same number s of columns, it suffices to show that they have the same null space, hence the same nullity q, for then (Theorem 3–9) both must have the same column rank $s - q$. If $AX = 0$, then $(PA)X = P(AX) = 0$, so that the null spaces $N(A)$ and $N(PA)$ of A and PA obey the inclusion,

$$N(A) \leqq N(PA).$$

By this very result

$$N(PA) \leqq N(P^{-1}PA) = N(A),$$

so that $N(A)$ both contains and is contained in $N(PA)$. This completes the proof by the comments made at the outset.

EXERCISES

1. Find a basis for the null space of the matrix

$$A = \begin{bmatrix} 3 & 1 & -1 \\ 0 & 1 & 2 \end{bmatrix}.$$

Then give a formula for all the solutions of the homogeneous system $AX = 0$.

2. State and prove the row analogue of Lemma 3–4.

3. The proof of Theorem 3–9 uses language which assumes that the null space of A is not the zero space. Give a separate proof of the theorem in the case in which this null space is zero.

3–9 Rank. The tools are now available for proving that the column rank of any matrix is equal to the row rank.

LEMMA 3–5. *If P and Q are nonsingular, PAQ has the same row rank as A and the same column rank as A.*

By Lemma 3–4 the column ranks of A and PA are the same, and by Lemma 3–3, part (ii), the column ranks of PA and PAQ are the same. This proves the "column part" of the lemma. The row space of $B = PAQ$ is the same as the column space of $B' = Q'A'P'$. Since Q' and P' are nonsingular [see (5)] the part of the lemma already proved implies that B' and A' have the same column ranks. But these equal numbers are the row ranks of B and A.

THEOREM 3–10. *The row rank of any matrix is equal to the column rank of the matrix.*

Let A be the given matrix. By Theorem 3–7 there exist matrices P and Q, which are products of elementary matrices, and thus are nonsingular, such that

$$(13) \qquad PAQ = \begin{bmatrix} I_r & 0 \\ 0 & 0 \end{bmatrix} = B.$$

Both the row rank and the column rank of B are equal to r. Then (Lemma 3–5) r is both the row rank and column rank of A, as claimed.

Although a matrix A may have one hundred columns and only five rows, its maximum number of linearly independent columns is the same as the corresponding number for rows by Theorem 3–10. This fact is only superficially surprising, since the hundred columns would belong to $V_5(\mathfrak{F})$, so that a linearly independent set of columns could include no more than five columns. For each matrix A the number characterized both as the column rank and the row rank of A is defined to be the *rank* of A.

For each rectangular matrix A an equivalent simplified matrix (13) is provided by Theorem 3–7, but we have not yet proved that B is unique. The integer r is the rank of B, and by Lemma 3–5 it is also the rank of A. Since r is thus completely determined by A, B is also, as shown by (13) and by the fact that B or any other matrix equivalent to A has the same size as A. This proves half of the next theorem.

THEOREM 3–11. *Each matrix A is equivalent to a unique matrix B of the type shown in (13), the integer r being the rank of A. Moreover,* *two $n \times s$ matrices over $\mathfrak{F}$ are equivalent if and only if they have the same rank.*

If A and C are $n \times s$ matrices with the same rank r, both are equivalent to the same matrix (13), hence are equivalent to each other. Conversely, if A and C are equivalent, $C = PAQ$ where P and Q are products of elementary matrices, hence nonsingular. Lemma 3–5 then implies that A and C have the same rank. This completes the proof of Theorem 3–11.

If A is a matrix over a field $\mathfrak{F}$, the rank of A has been defined as the dimension of a certain vector space over $\mathfrak{F}$, namely, the row space of A. However, A is also a matrix over $\mathfrak{K}$ for every field $\mathfrak{K}$ containing $\mathfrak{F}$, and the question arises whether A has the same rank over $\mathfrak{K}$ as it has over $\mathfrak{F}$.

COROLLARY 3–11. *The rank of a matrix A over $\mathfrak{F}$ does not change when the field of scalars is enlarged.*

Suppose that over $\mathfrak{F}$ the matrix A is equivalent to B in (13), and that by use of elementary operations over the larger field $\mathfrak{K}$ the matrix A can be transformed to a matrix B_1 of the type (13) but with r_1 ones on the diagonal. Then over the scalar field $\mathfrak{K}$, A is equivalent to both B and B_1, and the uniqueness in Theorem 3–11 implies that $B = B_1$, and thus that $r = r_1$.

EXERCISES

★1. Let $\xi_1, \ldots, \xi_s$ be linearly independent vectors belonging to $V_n(\mathfrak{F})$, and let $\mathfrak{K}$ be a field containing $\mathfrak{F}$. Prove that $\xi_1, \ldots, \xi_s$ are linearly independent as vectors velonging to $V_n(\mathfrak{K})$.

2. Compute the ranks of the matrices

$$\begin{bmatrix} 2 & 3 & 4 & 5 & 10 \\ -1 & -2 & 0 & 1 & 7 \end{bmatrix}, \quad \begin{bmatrix} 1 & 5 & 3 \\ -1 & 3 & 5 \\ 1 & 0 & -2 \end{bmatrix},$$

(a) by examining the number of linearly independent rows; (b) by examining the number of linearly independent columns; (c) by reducing each matrix to an equivalent matrix of the form (13).

★3. Show that the nullity of a matrix does not change when the scalar field is enlarged.

4. Prove that every matrix of rank r is a sum of r matrices of rank 1.

5. Let A be an $n \times s$ matrix of rank $r > 0$. Prove that there are matrices B and C such that B is $n \times r$, C is $r \times s$, and $A = BC$. Show that B and C are not unique but that both B and C must have rank r. Hint: A has r linearly independent rows.

6. Show that A and A' always have the same rank.

★7. Show that the rank of AB is no greater than the rank of either factor.

8. Show that the nullity of AB is at least as great as the nullity of B, and is at least as great as the nullity of either factor if B is square. Hint: Use Exercise 7.

3–10 Canonical sets. A set of $n \times s$ matrices is a *canonical set under equivalence* if every $n \times s$ matrix is equivalent to one and only one matrix in the set. Theorem 3–11 provides such a set.

The relation of equivalence separates the class of all $n \times s$ matrices over $\mathfrak{F}$ into subclasses, each subclass $\{A\}$ consisting of all $n \times s$ matrices equivalent to some given matrix A. A canonical set is then constructed by choosing in any way whatsoever a single matrix from each subclass. Theorem 3–11 points out a particularly simple matrix to choose from each subclass. Moreover, it points out that each subclass is distinguished by a single integer, the rank, common to all members of the subclass.

For $n \times n$ matrices the number of matrices in any canonical set under equivalence is $n + 1$. Why? What is the number for $n \times s$ matrices?

A canonical set for row equivalence is provided by Theorem 3–3. What remains to be done to complete the proof of this assertion?

3–11 Left and right inverses. An $n \times s$ matrix A is said to have B as a *left inverse* if $BA = I$. In this case I must be $s \times s$, and B then must be $s \times n$. Similarly, if there is a matrix C such that $AC = I$, then C is called a *right inverse* of A. The matrix I in this case must be $n \times n$ and C must be $s \times n$.

LEMMA 3–6. *If A has both a left inverse B and a right inverse C, A is nonsingular and $B = C = A^{-1}$.*

For proof we simply form the product BAC in the two ways permitted by the associative law:

$$(BA)C = I_s C = C = B(AC) = BI_n = B.$$

Thus $B = C$, so that this matrix is a "two-sided" inverse of A, and is therefore its unique inverse A^{-1}.

LEMMA 3–7. *Let A be an $n \times s$ matrix. Then*

(a) *A has a left inverse if and only if the rank of A is s.*

(b) *A has a right inverse if and only if the rank of A is n.*

To prove the necessity of condition (a), suppose that B is a left inverse: $BA = I_s$. The rows of I_s lie in the row space of A (Lemma 3-2) and are linearly independent. Hence the row rank r of A is at least s: $r \geqq s$. The column rank is also r, so A must have r linearly independent columns, and since there are only s columns altogether, $r \leqq s$, whence $r = s$.

Conversely, suppose that A has rank s. Then A has s linearly independent rows

$$A_{c_1}, A_{c_2}, \ldots, A_{c_s},$$

all belonging to $V_s(\mathfrak{F})$, hence forming a basis of $V_s(\mathfrak{F})$. For simplicity we write

$$\xi_j = A_{c_j}. \qquad\qquad (j = 1, \ldots, s)$$

Then all vectors in $V_s(\mathfrak{F})$, in particular the unit vectors u_i, are linear combinations of $\xi_1, \ldots, \xi_s$, and therefore

$$(14) \qquad\qquad u_i = b_{i1}\xi_1 + \cdots + b_{is}\xi_s \qquad\qquad (i = 1, \ldots, s)$$

for suitable scalars b_{ij}. To construct, one row at a time, a matrix B functioning as a left inverse of A, recall first that B must be $s \times n$. Then use in each row i the elements $b_{i1}, \ldots, b_{is}$ located in columns $c_1, \ldots, c_s$, respectively. All other positions in each row i are to be filled by zeros. Then row i of BA is given by (14), which is row i of I_s. This proves that the matrix B so constructed has the requisite property.

To prove (b) one may follow the pattern given for (a). A simpler method is to use transposes, taking into consideration the fact that A and A' have the same rank. Suppose A has rank n. The $s \times n$ matrix A' has rank n (its number of columns) so that by (a) there is a left inverse for A'. There is no loss of generality if this inverse is denoted by B':

$$B'A' = I_n.$$

Taking transposes now gives $AB = I_n$, and establishes half of (b). The converse is an easy reversal of these steps.

One-sided inverses may not be unique. Example:

$$A = \begin{bmatrix} I_s \\ 0 \end{bmatrix}, \quad B = (I_s, D).$$

Then $BA = I_s$ although the block D is arbitrary.

1. Prove the converse of (b) in Lemma 3–7, proof of which was not given in detail above.

2. If A has a left inverse B, prove that all left inverses of A are given by the formula $B + N$, N varying over all matrices (of the same size as B) such that $NA = 0$.

3. State and prove the analogue of Exercise 2 for right inverses.

4. Determine the existence of a left or a right inverse for each of the following matrices over $\mathfrak{R}$:

$$\begin{bmatrix} 2 & 1 \\ 5 & 2 \end{bmatrix}, \quad \begin{bmatrix} 2 & 1 & 3 \\ 5 & 2 & 0 \end{bmatrix}, \quad (1, 2, 3).$$

5. If an $n \times s$ matrix A has a left inverse but not a right inverse, show that $n > s$. Then show that there is an $s \times n$ matrix $N \neq 0$ such that $NA = 0$. Hint: First find a vector $\xi \neq 0$ such that $\xi A = 0$.

6. Use Exercises 2 and 5 to show that if a singular matrix has a left inverse, it has more than one left inverse. Hence, if A has a unique left inverse, it is nonsingular.

7. Show that A has a left inverse if and only if the null space of A is the zero space, and A has a right inverse if and only if the null space of A' is zero.

3–12 Nonsingular matrices. There are numerous tests for nonsingularity of a matrix. We begin with a theorem which is an easy by-product of the last section.

THEOREM 3–12. *If a matrix A has any two of the properties* (a), (b), *and* (c), *it is nonsingular:*
  (a) A *has a left inverse.*
  (b) A *has a right inverse.*
  (c) A *is square.*

If properties (a) and (b) are assumed, the conclusion is provided by Lemma 3–6. If (c) and (a) are assumed, A is $n \times n$ and (a) of Lemma 3–7 shows that A must have rank n. By (b) of the lemma, A has a right inverse as well as a left inverse, hence is nonsingular. If (c) and (b) are assumed the proof is similar to the previous case.

In contrast to matrices which are not square, if a square matrix A has a "one-sided" inverse, say a left inverse B, it follows from Theorem 3–12 that B is unique and in fact $B = A^{-1}$.

THEOREM 3–13. *The following three properties of a matrix A are equivalent:*
  (a) A *is nonsingular.*

(b) *A is square and its nullity is* 0.

(c) *A is* $n \times n$ *and its rank is* n.

Since (Theorem 3–9)

$$\text{nullity} = n - \text{rank},$$

it is clear that the nullity is 0 if and only if the rank is n, so (b) and (c) are equivalent. The proof will then be complete if we show that (a) and (c) are equivalent. For this we refer to Lemma 3–7. If A is nonsingular it is $n \times n$ and by this lemma its rank is n. If A is $n \times n$ and has rank n, the same lemma shows that A has both a left inverse and a right inverse, whence (Lemma 3–6) A is nonsingular.

THEOREM 3–14. *A matrix is nonsingular if and only if it is a product of elementary matrices. Two matrices B and A are equivalent if and only if $B = PAQ$, where P and Q are nonsingular.*

If M is a product $E_1 E_2 \cdots E_t$ of elementary matrices E_i, then each E_i has an inverse, and the matrix

(15) $$L = E_t^{-1} \cdots E_2^{-1} E_1^{-1}$$

satisfies the equations, $LM = I = ML$. Thus M is nonsingular. Conversely, if M is nonsingular it is $n \times n$ and has rank n. By Theorem 3–11 M is equivalent to I, $PMQ = I$,* where P and Q are products of elementary matrices. Then

(16) $$M = P^{-1} Q^{-1}.$$

As (15) indicates, the inverse of a product of elementary matrices is a product of elementary matrices. Then P^{-1}, Q^{-1}, and thus M, are products of elementary matrices. The final statement is an obvious application of the first statement to the criterion we have been using for equivalence.

THEOREM 3–15. *A matrix A is nonsingular if and only if it is row equivalent to the identity matrix.*

We shall refer to the matrix B of Theorem 3–3, which is row equivalent to A. This matrix B has r nonzero rows, where r is the rank of A. If A is nonsingular it is $n \times n$ and has rank n, so that $r = n$. Then the r column numbers c_i obeying requirement (c) of Theorem 3–3 must be

$$c_1 = 1, \quad c_2 = 2, \ldots, \quad c_n = n.$$

---

* We could express this fact by the equation $M = PIQ = PQ$ and thereby shorten the argument.

Property (d) of Theorem 3–3 then implies that $B = I_n$. Conversely, if A is row equivalent to I_n, we have $PA = I_n$. Then A is $n \times n$ and has a left inverse, so that (Theorem 3–12) A is nonsingular. Another simple proof of this theorem can be made by a direct application of the preceding theorem.

This result is rather surprising, since many matrices cannot be reduced to their equivalent canonical matrices of Theorem 3–11 by row operations alone. An example:

$$A = \begin{bmatrix} 1 & 1 \\ 1 & 1 \end{bmatrix}, \quad B = \begin{bmatrix} 1 & 0 \\ 0 & 0 \end{bmatrix}.$$

Since A has rank 1 its simplified matrix of Theorem 3–11 is the matrix B shown. The first row of B is not a linear combination of the rows of A; therefore B is not obtainable by row operations alone.

The property in the last theorem furnishes one of the simplest ways to compute the inverse of a matrix A which has an inverse. There must be elementary matrices E_i such that

$$I = E_t \cdots E_1 A.$$

Multiplying on the right by A^{-1} gives

$$A^{-1} = E_t \cdots E_1 = (E_t \cdots E_1)I.$$

This equation is the clue to the method. Multiplying $E_1, \ldots, E_t$ in turn onto I may be interpreted as performing row operations on I, the same row operations performed on A in reducing it to the identity. This proves

THEOREM 3–16. *If A is reduced to the identity matrix I by a succession of row operations, the same succession of row operations performed on I produces A^{-1}.*

EXERCISES

1. If a square matrix A has a left inverse B, prove that it has only one left inverse.

2. Show that an $n \times n$ matrix over $\mathfrak{F}$ is nonsingular if and only if its rows (or columns) form a basis of $V_n(\mathfrak{F})$.

3. Prove Theorem 3–15 as a corollary of Theorem 3–14.

★4. Let A be $n \times n$. Show that A is nonsingular if and only if $A\xi \neq A\eta$ for all pairs of distinct column vectors ξ and η belonging to $V_n(\mathfrak{F})$.

5. Use Theorem 3–16 to compute the inverses of the following real matrices, then check each inverse by multiplication:

(a) $\begin{bmatrix} 2 & 1 \\ 3 & 2 \end{bmatrix}$;　　　　　　(b) $\begin{bmatrix} 2 & 1 \\ 5 & 2 \end{bmatrix}$;

(c) $\begin{bmatrix} 0 & 1 & 2 \\ -1 & 3 & 0 \\ 1 & -2 & 1 \end{bmatrix}$;　　　(d) $\begin{bmatrix} 2 & -1 & 3 \\ 1 & 0 & -1 \\ 0 & -2 & 1 \end{bmatrix}$.

6. Find the inverse of the following matrix of rational functions of x:

$$A = \begin{bmatrix} x & \dfrac{1}{x+1} \\[2ex] 1 & \dfrac{2}{x(x+1)} \end{bmatrix}.$$

7. Let y_1 and y_2 be real functions of a real variable x, and let their derivatives with respect to x be denoted by y_1' and y_2'. Use Exercise 6 to find y_1 and y_2 if they satisfy the system of differential equations,

$$xy_1' + \frac{1}{x+1}\, y_2' = 6x^2,$$

$$y_1' + \frac{2}{x(x+1)}\, y_2' = 6(x-1).$$

★8. If a product $Q_1 \cdots Q_r$ of square matrices Q_i is nonsingular, prove that every Q_i is nonsingular.

3–13 All bases of a vector space. We have seen that every vector space V has a basis, and that if V has dimension s a basis is an arbitrary set of s linearly independent vectors in V. In terms of nonsingular matrices we shall see how to determine all bases from a given one.

First consider $V_n(\mathfrak{F})$. If the vectors in a basis are written as the columns of a matrix A, this matrix is $n \times n$ and has rank n, since its columns are linearly independent. Theorem 3–13 then implies that A is nonsingular. Conversely, if A is an $n \times n$ nonsingular matrix over $\mathfrak{F}$, its columns form a basis of $V_n(\mathfrak{F})$. Clearly, bases of $V_n(\mathfrak{F})$ can also be characterized in terms of rows of nonsingular $n \times n$ matrices over $\mathfrak{F}$.

Given a second basis of $V_n(\mathfrak{F})$, its vectors may be used as columns of a nonsingular matrix B. Then there is a nonsingular matrix P over $\mathfrak{F}$ such that

$$(17) \qquad\qquad B = AP,$$

namely, $P = A^{-1}B$. Conversely, if an $n \times n$ nonsingular matrix P over $\mathfrak{F}$ is given, and if one basis of $V_n(\mathfrak{F})$ is given as the set of columns of a matrix A, then (17) determines a matrix B which is nonsingular, whence its columns provide a second basis.

These simple observations may be generalized to apply to any subspace V of dimension s of $V_n(\mathfrak{F})$. A basis of V may be written as the columns of an $n \times s$ matrix A over $\mathfrak{F}$, the rank of A being s. (Why?) If P is an $s \times s$ nonsingular matrix over $\mathfrak{F}$, the matrix

$$(18) \qquad B = AP$$

is $n \times s$ and (the column analogue of Lemma 3–2!) its column space coincides with that of A. Since this column space is V and has dimension s, the s columns of B form a basis of V. This shows how a second basis of V may be constructed from a fixed basis and a nonsingular $s \times s$ matrix P.

Conversely, given the second basis (as well as the first) there are $n \times s$ matrices B and A, whose columns constitute the new and old bases, respectively. We shall show that B and A are related by an equation (18) where P is nonsingular. The jth column vector η_j in B is a vector belonging to V, hence is a linear combination of the columns $\xi_1, \ldots, \xi_s$ of A:

$$(19) \qquad \eta_j = \xi_1 c_{1j} + \xi_2 c_{2j} + \cdots + \xi_s c_{sj}.$$

Since an equation (19) holds for each $j = 1, \ldots, s$, it follows that $B = AP$, where $P = (c_{ij})$, an $s \times s$ matrix. The roles of A and B in this argument may be interchanged, so that $A = BQ$, where Q is $s \times s$. Substitution of AP for B then gives

$$A = A(PQ),$$

where PQ is an $s \times s$ matrix over $\mathfrak{F}$. This equation implies that $PQ = I_s$. (But this is not trivial! The proof requires the fact that the columns of A form a basis of V. Why?)

Since P is square and has a right inverse, it is nonsingular. This completes the proof that $B = AP$ with P nonsingular.

EXERCISES

1. In the proof above show why $PQ = I_s$.

2. Let A be an $n \times s$ matrix over $\mathfrak{F}$ and let V be a subspace of dimension s of $V_n(\mathfrak{F})$. Show that the columns of A form a basis of V if and only if these columns belong to V, and A has a left inverse.

3. Formulate the main result of this section as a theorem. Then reformulate it as a true theorem in which rows of matrices are used as columns are used in the text.

3–14 Bilinear forms. If $x_1, \ldots, x_n$ and $y_1, \ldots, y_s$ are two sets of variables, a quadratic polynomial of the type

$$(20) \qquad f = \sum_{j=1}^{s} \sum_{i=1}^{n} x_i a_{ij} y_j$$

is called a *bilinear form* in these sets of variables. It can be written as a linear combination of the y_j,

$$f = f_1 y_1 + \cdots + f_s y_s, \quad f_j = \sum_i a_{ij} x_i,$$

with coefficients f_j, which are linear combinations of the x_i, and also as $f = g_1 x_1 + \cdots + g_n x_n$ with coefficients $g_i = \Sigma a_{ij} y_j$. If $X =$ col $(x_1, \ldots, x_n)$, $Y =$ col $(y_1, \ldots, y_s)$, and $A = (a_{ij})$, f is most conveniently written as the 1×1 matrix,

$$(21) \qquad f = X'AY,$$

where f has been written instead of (f).

Bilinear forms are often reduced or simplified by linear substitutions

$$(22) \qquad \begin{aligned} x_i &= p_{i1} t_1 + \cdots + p_{in} t_n, & (i = 1, \ldots, n) \\ y_j &= q_{j1} v_1 + \cdots + q_{js} v_s. & (j = 1, \ldots, s) \end{aligned}$$

To discuss these substitutions in matrix notation let

$$\begin{aligned} T &= \text{col } (t_1, \ldots, t_n), & V &= \text{col } (v_1, \ldots, v_s), \\ P' &= (p_{ih}), & Q &= (q_{jk}). \end{aligned}$$

Although it may seem peculiar to indicate the coefficient matrix (p_{ih}) as the transpose of another matrix P, this will simplify our final formula. Substitutions (22) may now be written as

$$(23) \qquad X = P'T, \qquad Y = QV.$$

When these are employed in (21) we obtain

$$\begin{aligned} f &= X'AY = T'PAQV = T'BV, \\ B &= PAQ. \end{aligned}$$

Thus the linear substitutions (22) in a bilinear form amount to replacing the matrix A of the form by PAQ, where P and Q are square.

Most often one is interested in substitutions which are reversible, that is, substitutions (22) which can be solved uniquely for the t's in terms of the x's, and for the v's in terms of the y's. This means that there are matrices R and S such that

$$T = RX, \qquad V = SY.$$

By (23) we find that

$$X = (P'R)X, \qquad Y = (QS)Y,$$

whence

$$P'R = I_n, \qquad QS = I_s.$$

The square matrices P' and Q then are nonsingular and so is P. Conversely, if P and Q are nonsingular, the substitutions (23) are reversible:

$$T = (P')^{-1}X, \quad V = Q^{-1}Y.$$

The matrix $A = (a_{ij})$ in (21) and (20) is called the *matrix of the bilinear form*. Apart from notation for the variables the form is completely determined by its matrix. The discussion just given may be summarized thus: Nonsingular linear substitutions in a bilinear form f with matrix A yield a new form with matrix PAQ equivalent to A. The problem of reducing f by nonsingular linear substitutions is therefore identical with the problem of reducing a rectangular matrix by equivalence. By Theorem 3–11 if f has rank r (that is, if the matrix A of f has rank r), there are nonsingular linear substitutions converting f to

$$t_1 v_1 + \cdots + t_r v_r.$$

EXERCISES

1. Formulate the matrix problem which is equivalent to the problem of finding nonsingular linear substitutions reducing the bilinear form

$$f = x_1 y_1 + 2x_1 y_2 + 3x_1 y_3 + x_2 y_2 - x_2 y_3$$

to

$$t_1 v_1 + t_2 v_2.$$

2. Solve the matrix problem referred to in Exercise 1. Then check by actual substitution.

3–15 Minimum polynomial and inverses. It was proved at the end of Chapter 2 that every square matrix satisfies a polynomial equation of the type

$$(24) \qquad A^t + k_{t-1}A^{t-1} + \cdots + k_1 A + k_0 I = 0.$$

It is convenient to say that A is a "root" of the polynomial

$$(25) \qquad g(x) = x^t + k_{t-1}x^{t-1} + \cdots + k_1 x + k_0,$$

meaning thereby that (24) is true. Note the insertion of I as a multiplier on the constant term k_0 of (25) when x is replaced by A.

The sum on the left side of (24) would have no meaning if the last term were a scalar. When A is a root of (25) we shall write $g(A) = 0$ to represent equation (24).

Among all nonconstant polynomials of the form (25) having A as a root there is one of minimal degree. We shall prove that this polynomial, called the *minimum function* or *minimum polynomial* of A, is unique.

THEOREM 3-17. *Each square matrix over $\mathfrak{F}$ has a unique minimum polynomial.*

Suppose that A were a root of polynomials

$$(26) \qquad \begin{aligned} m(x) &= x^e + c_{e-1}x^{e-1} + \cdots + c_1 x + c_0, \\ p(x) &= x^e + d_{e-1}x^{e-1} + \cdots + d_1 x + d_0, \end{aligned}$$

of the same minimal degree. Then A would be a root of

$$(27) \qquad m(x) - p(x) = (c_{e-1} - d_{e-1})x^{e-1} + \cdots + (c_0 - d_0).$$

If any term on the right side of (27) has nonzero coefficient $c_i - d_i$, let $c_j - d_j$ be the first such coefficient. Multiplying by the inverse of $c_j - d_j$ we find a polynomial of the type (25) having A as a root and having degree less than e, in defiance of the definition of e as the degree of the minimum polynomial. This proves that all $c_i - d_i = 0$ and every $c_i = d_i$, whence $p(x) = m(x)$.

A nonsquare matrix A has no minimum polynomial, since powers A^2, A^3, etc., are not defined.

The minimum polynomial of a square matrix A may be used to determine whether A has an inverse, and even to calculate the inverse when it exists.

THEOREM 3-18. *A square matrix has an inverse if and only if the constant term of its minimum polynomial is nonzero.*

Let (26) be the minimum polynomial of A, and suppose that $c_0 \neq 0$. In the equation

$$(28) \qquad A^e + c_{e-1}A^{e-1} + \cdots + c_1 A + c_0 I = 0$$

we may subtract $c_0 I$ from both sides, then multiply by $(-c_0)^{-1}$, obtaining

$$b_e A^e + \cdots + b_1 A = I,$$

where $b_e = (-c_0)^{-1}$ and $b_i = -c_i/c_0$ for $i = 1, \ldots, e - 1$. This equation may be factored:

$$A(b_e A^{e-1} + \cdots + b_1 I) = I = (b_e A^{e-1} + \cdots + b_1 I)A.$$

The matrix in parentheses is the inverse of A.

Conversely, suppose that A has an inverse. If $c_0 = 0$, equation (26) becomes

$$A(A^{e-1} + c_{e-1}A^{e-2} + \cdots + c_1 I) = 0.$$

Multiplying by A^{-1} on the left gives

$$A^{e-1} + \cdots + c_1 I = 0,$$

an equation which conflicts with the fact that e is the degree of the minimum polynomial of A. Hence $c_0 \neq 0$.

EXERCISES

1. Find the minimum polynomial of the matrix

$$\begin{bmatrix} 0 & 1 \\ 0 & 0 \end{bmatrix}.$$

2. Find the minimum polynomial $m(x)$ of the real matrix

$$A = \begin{bmatrix} 1 & 2 \\ 3 & 4 \end{bmatrix},$$

given that $m(x)$ has degree 2. If possible, use $m(x)$ to calculate A^{-1}.

3. Given that the matrix

$$A = \begin{bmatrix} 0 & 1 & 0 \\ 0 & 0 & 1 \\ c & b & a \end{bmatrix}$$

has minimum polynomial $m(x)$ of degree 3, find $m(x)$. Then find a necessary and sufficient condition on a, b, and c for A to be nonsingular. Assuming this condition to hold, find a formula for A^{-1} as a quadratic polynomial in A.

4. If A is square, prove the equivalence of the following four properties: (a) A is singular. (b) There is a square matrix $B \neq 0$ such that $BA = 0$. (c) There is a square matrix $C \neq 0$ such that $AC = 0$. (d) The minimum polynomial of A has zero constant term.

5. Find the minimum polynomial of the real matrix $A = \text{diag}\,(1, 1, 1, 2, 3)$. Then find the minimum polynomial of an arbitrary diagonal matrix $A = \text{diag}\,(a_1, \ldots, a_n)$.

CHAPTER 4

DETERMINANTS

4–1 Definition. The use of determinants in the solution of systems of two linear equations in two unknowns is familiar to most students of algebra. Although determinants are not important for this purpose, they do play a significant role in many theoretical investigations. We shall define the determinant of a general square matrix.

Let $A = (a_{hj})$ be an $n \times n$ matrix, and consider products

$$(1) \qquad a_{1i_1}a_{2i_2} \cdots a_{ni_n},$$

in which there are n factors, one from each row of A. The second subscripts

$$(2) \qquad i_1, i_2, \ldots, i_n$$

in (1) denote some arrangement of

$$(3) \qquad 1, 2, \ldots, n,$$

and (2) is called a *permutation* of (3). The number of such permutations is $n! = n(n-1) \cdots 1$, and this is the number of products of the type (1) which can be formed from the given matrix A. Notice that in each product (1) precisely one factor comes from each column and one from each row.

With each product (1) constructed from the matrix A we shall associate a factor which is plus or minus one, that is, a power $(-1)^t$. This "sign" is determined by the permutation (2) in a manner that will be described presently. Then the sum of all terms

$$(4) \qquad (-1)^t a_{1i_1}a_{2i_2} \cdots a_{ni_n}$$

is called the *determinant* of A and is denoted by $|A|$.

These ideas are borne out by the familiar formula

$$\begin{vmatrix} a_{11} & a_{12} \\ a_{21} & a_{22} \end{vmatrix} = a_{11}a_{22} - a_{12}a_{21}$$

for the determinant of a 2×2 matrix A. The two terms are all possible terms (1) that can be formed with $n = 2$. One term has the sequence 1, 2 of second subscripts, and has been multiplied by $(-1)^t$ = 1; the other has the sequence 2, 1 of second subscripts and has been.

multiplied by $(-1)^t = -1$. Notice that the first subscripts must be in natural order before we take down the sequence of second subscripts. Section 4–2 shows how to use this sequence to determine the signs $(-1)^t$.

4–2 Transpositions. To discuss the sign $(-1)^t$, which is determined separately for each term (4), we consider ways of restoring the sequence (2) to the natural order (3) by *transpositions*, that is, interchanges of only two integers at a time. For example, the permutation

(5)   1, 4, 3, 2

can be restored to natural order by the following three steps, each of which interchanges only one pair of integers:

(a) 1, 4, 2, 3,
(b) 1, 2, 4, 3,
(c) 1, 2, 3, 4.

We write $t = 3$ to denote the number of steps, or transpositions, which have been used. Then $(-1)^t = (-1)^3 = -1$ is the sign associated with the permutation 1, 4, 3, 2.

The restoration of (5) to natural order can be accomplished in other ways; for example, by interchanging the 4 and the 2. This employs only one transposition, so $t = 1$. Although this differs from the first value of t, the sign is the same:

$$(-1)^3 = (-1)^1 = -1.$$

Many other values of t can be found, but for the permutation (5) all of these values will be odd. This is part of a general rule:

LEMMA 4–1. *The number of transpositions required to restore a permutation (2) to natural order is not unique, but for a given permutation it is always even or always odd.*

To prove this lemma consider the following polynomial in variables $x_1, \ldots, x_n$:

$$P = \prod_{i<j} (x_i - x_j)$$
$$= (x_1 - x_2) \cdots (x_1 - x_n) \cdot (x_2 - x_3) \cdots (x_2 - x_n) \cdots (x_{n-1} - x_n).$$

Notice that the first subscript in each factor $x_i - x_j$ is less than the second. We shall study the effect on P of a transposition, interchanging only h and k, applied to the subscripts of the variables in P.

That is, we shall replace x_h by x_k, and x_k by x_h, wherever they occur in P.

Let Q denote the product of those factors $x_i - x_j$ of P involving neither x_h nor x_k. Then Q will not be affected by the proposed transposition. There will be one factor, $x_h - x_k$ or $x_k - x_h$, involving both h and k as subscripts, and it can be written as $\pm(x_h - x_k)$. Each remaining factor involves x_h or x_k, but not both. The product of those involving only x_h can be written as $\pm P_h$, and the product of those involving only x_k can be written as $\pm P_k$, where

$$P_h = \prod_{i \neq h,\, k} (x_i - x_h), \quad P_k = \prod_{j \neq h,\, k} (x_j - x_k).$$

Then

(6) $$P = \pm Q \cdot P_h \cdot P_k \cdot (x_h - x_k).$$

For example, take $n = 4$, $h = 3$, $k = 2$:

$$
\begin{aligned}
P &= (x_1 - x_2)(x_1 - x_3)(x_1 - x_4)(x_2 - x_3)(x_2 - x_4)(x_3 - x_4), \\
Q &= x_1 - x_4, \\
P_3 &= (x_1 - x_h)(x_4 - x_h) = -(x_1 - x_3)(x_3 - x_4), \\
P_2 &= (x_1 - x_k)(x_4 - x_k) = -(x_1 - x_2)(x_2 - x_4), \\
P &= Q \cdot (-P_3) \cdot (-P_2)(-1)(x_3 - x_2), \\
P &= -Q \cdot P_3 \cdot P_2(x_3 - x_2).
\end{aligned}
$$

From (6) it is easy to see the effect of the proposed transposition: Q is not changed at all; $P_h P_k$ is replaced by $P_k P_h$; and $x_h - x_k$ is replaced by its negative, $x_k - x_h$. Altogether, the transposition replaces P by $-P$. If the restoration of (2) to the natural order (3) can be achieved with t transpositions, P will be converted to $(-1)^t P$.

The entire permutation from (2) to (3) can be carried out on the subscripts of the variables in P at a single stroke: Replace x_{i_1} everywhere in P by x_1; x_{i_2} by x_2; etc., and x_{i_n} by x_n. The result is a well-defined polynomial which may be indicated by P_0. If the same effect, however, can be achieved by t transpositions and also by s transpositions, we find that

$$(-1)^t P = P_0 = (-1)^s P.$$

It follows that $(-1)^t = (-1)^s$, as the lemma claims.

By virtue of this lemma, the sign or factor $(-1)^t$ associated with each term (1) of $|A|$ is uniquely determined, and the definition of $|A|$ is now complete.

72 DETERMINANTS [CHAP. 4

EXERCISES

1. Prove that $|A| = a_{11}a_{22} - a_{12}a_{21}$ if $A = (a_{ij})$ is 2×2.
2. Evaluate $|A|$, where $A = (a_{ij})$ is a general 3×3 matrix.
3. Compute the following determinants:

$$\begin{vmatrix} 1 & 2 & 3 \\ 0 & -4 & 2 \\ -1 & 5 & 4 \end{vmatrix}; \quad \begin{vmatrix} 0 & -1 & 1 & 4 \\ 3 & 2 & -2 & 1 \\ 0 & 4 & 0 & 1 \\ 1 & 0 & -1 & 1 \end{vmatrix}.$$

4-3 Special cases. For the identity matrix the definition of determinant produces only one term which is not zero, and this is 1:

$$|I| = 1.$$

In fact, the same reasoning shows that if D is any diagonal matrix, $|D|$ is the product of the diagonal elements:

(7) $$D = \text{diag } (a_1, \ldots, a_n), \quad |D| = a_1 a_2 \cdots a_n.$$

Now consider a matrix

(8) $$A = \begin{bmatrix} B_1 & 0 \\ 0 & B_2 \end{bmatrix},$$

in which B_1 is an $r \times r$ block, B_2 is an $s \times s$ block, and the 0's are rectangular zero matrices of suitable sizes. Then, as in Section 1-6, A is called the direct sum of B_1 and B_2, and the following notation is used:

$$A = \text{diag } (B_1, B_2).$$

Similarly, if there are t square blocks $B_1, \ldots, B_t$ on the diagonal, and blocks off the diagonal are all zero, we write

(9) $$A = \text{diag } (B_1, \ldots, B_t)$$

and call A the direct sum of $B_1, \ldots, B_t$. This may be regarded as a generalization of the concept of a diagonal matrix. The formula (7) for the determinant of a diagonal matrix also generalizes: The matrix A in (9) has determinant

(10) $$|A| = |B_1| \cdots |B_t|.$$

To prove (10) consider first the special case (8) in which $t = 2$. Let B_1 be $r \times r$ and B_2 be $s \times s$. Each term (4) of $|A|$ has first r factors from the first r rows of A. Any one of these factors that does not come from the first r columns is 0 and the term is 0. Thus the terms of $|A|$ coincide with the products of those of $|B_1|$ by those of

$|B_2|$. It is easy to see that the sign $(-1)^t$ of each term ot $|A|$ is the product of the signs $(-1)^{t_1}$ and $(-1)^{t_2}$ of the corresponding terms of $|B_1|$ and $|B_2|$. When these details are verified, we have formula (10) for the case $t = 2$. The general case follows easily from this special case:

$$A = \text{diag } (B_1, A_1), \quad A_1 = \text{diag } (B_2, \ldots, B_t),$$
$$|A| = |B_1| \cdot |A_1|, \quad |A_1| = |B_2| \cdot |\text{diag } (B_3, \ldots, B_t)|.$$

Thus repeated application of the result for $t = 2$ gives (10).

In this discussion we have been considering determinants of square diagonal blocks. The determinant of any square submatrix of A is called a *subdeterminant* (or a *minor*) of A and of $|A|$.

EXERCISES

★1. Let $A = (a_{ij})$ be an $n \times n$ matrix such that $a_{ij} = 0$ if $j > i$, that is, elements above the diagonal are zero. Show that $|A|$ is the product of the diagonal elements of A.

★2. Let A be a matrix of blocks with only zero blocks above the diagonal, and square blocks on the diagonal. Verify that $|A|$ is the product of the determinants of the diagonal blocks.

4-4 Elementary operations. The relations between $|A|$ and $|B|$ when B is equivalent to A will be studied now.

THEOREM 4-1. *If B is obtained from A by interchanging two rows,* $|B| = -|A|$.

Suppose that rows p and q are interchanged. Each term

$$(11) \qquad T = b_{1i_1} \cdots b_{ni_n}$$

in $|B|$, with sign $(-1)^t$ neglected, has factors $b_{hj} = a_{hj}$ except for

$$(12) \qquad b_{pi_p} = a_{qi_p}, \quad b_{qi_q} = a_{pi_q}.$$

Thus if the pth and qth factors in (11) are interchanged, (11) does not change in value but does acquire the appearance of a typical term S in $|A|$ with row subscripts in natural order. To restore the column subscripts of S to natural order, first restore them to the order in which they appear in (11), using one transposition. Then use the same t transpositions used on (11) as a term of $|B|$. This shows that if (11) as a term T of $|B|$ requires the sign $(-1)^t$, when it is written as a term of $|A|$ having the same value as T, it requires the sign $(-1)^{t+1} = -(-1)^t$. Since each term changes sign, $|B| = -|A|$.

THEOREM 4–2. *If B is obtained from A by multiplying a row by c,* $|B| = c \cdot |A|$.

The terms of $|B|$ are now the same as those of $|A|$, signs included, except for the replacement of the factors

(13)
$$a_{p1}, \ldots, a_{pn}$$

by $ca_{p1}, \ldots, ca_{pn}$. Since one and only one of the quantities (13) appears as a factor of each term of $|A|$, this replacement introduces the factor c on every term.

THEOREM 4–3. *If A' is the transpose of A,* $|A'| = |A|$.

Let $A = (a_{hj})$, $A' = (b_{hj})$ so that $b_{hj} = a_{jh}$. A typical term in $|A'|$ is

(14)
$$b_{1i_1} \cdots b_{ni_n} = a_{i_1 1} \cdots a_{i_n n}.$$

Then apart from associated signs, the terms comprising $|A'|$ are the same as those in $|A|$, since from either viewpoint (14) gives all possible products of n factors, one from each row and one from each column. Suppose that t transpositions restore $i_1, \ldots, i_n$ to natural order so that (14), regarded as a term T in $|B|$, has sign $(-1)^t$. If the right side of (14) is to be regarded as a term S of $|A|$ the factors must first be rearranged to get the row subscripts in order before we can count the transpositions that determine the sign associated with S. This rearrangement of *factors* on the right side of (14) is exactly parallel to the rearrangement of *column subscripts* on the left of (14), hence may be carried out in t steps, each step interchanging only two factors. If the result is denoted by

(15)
$$a_{1j_1} \cdots a_{nj_n},$$

the arrangement

(16)
$$j_1, \ldots, j_n$$

has been obtained from $1, \ldots, n$ by t transpositions. The same transpositions performed in reverse order then restore (16) to natural order. Thus (15) does have the same sign $(-1)^t$ as (14), and the theorem is established.

By virtue of Theorem 4–3, results relating to rows can be converted to analogues for columns. Let B denote the result of interchanging column p and column q in A. Then B' is the result of interchanging row p and row q in A', so that $|B'| = -|A'|$, whence by Theorem 4–3, $|B| = -|A|$. Similarly Theorem 4–2 may be converted to a column theorem: $|B| = c \cdot |A|$ if B results from multiplying a column of A by c.

THEOREM 4–4. *The column analogues of Theorems 4–1 and 4–2 are valid.*

EXERCISES

1. Verify Theorem 4–3 for a general 2×2 matrix.

2. Show that

$$\begin{vmatrix} 2 & 4 & 6 \\ 0 & 2 & 3 \\ 1 & 4 & 9 \end{vmatrix} = 12 \begin{vmatrix} 1 & 1 & 1 \\ 0 & 1 & 1 \\ 1 & 2 & 3 \end{vmatrix} = 12.$$

3. Let $A = (a_{ij})$ be an $n \times n$ real matrix such that every diagonal element is zero and $a_{ij} = -a_{ji}$ if $i \neq j$. Prove that if n is odd, $|A| = 0$. Hint: Use Theorems 4–2, 4–3.

★4. Let A be a square, complex matrix, and $\bar{A}$ be the matrix obtained from A by replacing each element of A by its complex conjugate. Show that $|\bar{A}|$ is the conjugate of $|A|$.

5. In the preceding exercise if $\bar{A} = A'$, prove that $|A|$ is real.

4–5 Expansion by cofactors. Before discussing the third type of elementary operation it is necessary to discuss another concept.

DEFINITION 1. Let M_{hj} denote the $(n-1) \times (n-1)$ submatrix of $A = (a_{hj})$ obtained by deleting row h and column j. Then the scalar

$$c_{hj} = (-1)^{h+j} |M_{hj}|$$

is called the *cofactor* of a_{hj} in A. The $n \times n$ matrix $(c_{hj})'$ is called the *adjoint* of A and is denoted by adj A.

Thus, if (c_{hj}) is called the cofactor matrix of A, the adjoint of A is the transpose of the cofactor matrix of A.

THEOREM 4–5. *For each value of $h = 1, 2, \ldots, n$,*

(17) $$|A| = a_{h1}c_{h1} + \cdots + a_{hn}c_{hn},$$
(18) $$|A| = c_{1h}a_{1h} + \cdots + c_{nh}a_{nh}.$$

Each term (4) of $|A|$ contains precisely one factor from row h,

$$a_{h1}, \ldots, a_{hn},$$

of A. If all the terms of $|A|$ containing a_{hj} are assembled, and if this is done for each j, the following type of formula results:

$$|A| = a_{h1}d_{h1} + \cdots + a_{hn}d_{hn}.$$

Then (17) amounts to the claim that $d_{hj} = c_{hj}$. Thus we must prove that

(19) $$d_{hj} = (-1)^{h+j}|M_{hj}|,$$

where M_{hj} is the submatrix of A that remains when row h and column j are deleted. Since d_{hj} is, apart from signs of the terms, the sum of all products of $n-1$ factors a_{pq} representing every row of A except row h and every column except column j, the terms in d_{hj} are the same as those in $|M_{hj}|$. It remains only to see that each term has the appropriate sign.

To take first the case in which $h = j = 1$, we must prove that $d_{11} = |M_{11}|$; that is, each term in d_{11} must have the same associated sign as the term in M_{11}, which is made up of the same factors. Each term of d_{11} is of the form

(20) $$(-1)^t a_{2i_2} \cdots a_{ni_n},$$

where

$$S = (-1)^t a_{11} a_{2i_2} \cdots a_{ni_n}$$

is a term of $|A|$. Since the sign $(-1)^t$ of S depends only on the permutation $i_2, \ldots, i_n$ of $2, \ldots, n$, (20) does have the appropriate sign for a term of $|M_{11}|$. Thus $d_{11} = |M_{11}|$. From this special case of (19) the general case may be proved.

Interchange row h of A with the row above it, and repeat this process until after $h-1$ interchanges row h appears in the top position. The lower rows then are in their normal order except for the omission of row h. Similarly, move column j to the first position by $j-1$ interchanges, leaving the remaining columns in their usual order, except for the omission of column j. The matrix $B = (b_{pq})$ so obtained results from $j - 1 + h - 1 = h + j - 2$ interchanges, so that

(21) $$|A| = (-1)^{h+j}|B|.$$

Moreover, the element b_{11} of B is $b_{11} = a_{hj}$, and the submatrix B_{11} obtained by deleting row 1 and column 1 of B is $B_{11} = M_{hj}$.

By the special case of (19) already proved, the sum of all terms in $|B|$ containing $b_{11}(= a_{hj})$ as a factor is $b_{11}|B_{11}|$, and this is equal to $a_{hj}|M_{hj}|$. By (21), when the terms of $|B|$ are multiplied by $(-1)^{h+j}$ they are converted to terms of $|A|$, so that the terms in

$$(-1)^{h+j} a_{hj}|M_{hj}|$$

are terms in $|A|$. This gives (19), and thus (17). The analogous formula (18) is obtained from (17) by an argument on transposes.

The formula (17) amounts to the statement that $|A|$ may be calculated by multiplying each element of row h by its cofactor and adding the results. This is called the *expansion* of $|A|$ by elements of row h. Similarly, (18) is the expansion of $|A|$ by elements of column h: $|A|$ is the sum of the products obtained by multiplying each element of column h by its cofactor. A determinant may be computed or "expanded" according to the elements of any row or any column.

EXERCISES

1. Expand each of the matrices in Exercise 3, Section 4-2 (a) according to the elements of column 1, (b) according to the elements of row 1.

2. If a square matrix has a row of zeros, prove that its determinant is zero, first by the definition of determinant, second by use of Theorem 4-5.

★3. Let A be an $n \times n$ matrix for which every $k \times k$ subdeterminant is 0 for some fixed $k < n$. Prove that $|A| = 0$.

4-6 The adjoint. If two rows of A are alike, interchanging them gives the same matrix A, but also gives $|A| = -|A|$ by Theorem 4-1. Then $(1 + 1)|A| = 0$, and $|A| = 0$. The corresponding result for columns is provable in the same way or by the device of transposes.

LEMMA 4-2. *If two rows or two columns of a square matrix A are alike, $|A| = 0$.*

There are certain fields in which $1 + 1 = 0$, whence the argument above fails: $(1 + 1)|A| = 0$, although $|A|$ may be different from zero. Even in this strange situation Lemma 4-2 is valid, but the proof will not be given here.

THEOREM 4-6. *For every square matrix A,*

$$A(\text{adj } A) = |A| \cdot I = (\text{adj } A)A.$$

In short, when a matrix A is multiplied on either side by its adjoint, a scalar matrix is produced, the diagonal elements all equaling $|A|$.

Column h of adj A consists of the cofactors of the elements $a_{h1}, \ldots,$ a_{hn}, of row h of A. Thus the element in the (h, h) place of $A(\text{adj } A)$ is

$$a_{h1}c_{h1} + \cdots + a_{hn}c_{hn},$$

which is $|A|$ by Theorem 4-5. To complete the proof of $A(\text{adj } A) = |A| \cdot I$, we must show that

$$(22) \qquad\qquad a_{j1}c_{h1} + \cdots + a_{jn}c_{hn} = 0,$$

when $j \neq h$. Suppose that row h of A is replaced by a vector identical with row j, giving a new matrix B. Then $|B| = 0$ by Lemma 4–2. If the determinant of B is expanded by the elements of row h, however, the left side of formula (22) results. This completes the proof that $A(\text{adj } A) = |A| \cdot I$. That $(\text{adj } A) \cdot A = |A| \cdot I$ can be proved similarly.

EXERCISES

1. Prove the column part of Lemma 4–2 from the row part by use of transposes.

★2. The matrix

$$V = \begin{bmatrix} 1 & x_1 & x_1^2 & \cdots & x_1^{n-1} \\ 1 & x_2 & x_2^2 & \cdots & x_2^{n-1} \\ \cdot & \cdot & \cdot & & \cdot \\ \cdot & \cdot & \cdot & & \cdot \\ \cdot & \cdot & \cdot & & \cdot \\ 1 & x_n & x_n^2 & \cdots & x_n^{n-1} \end{bmatrix}$$

is known as Vandermonde's matrix. Prove that $|V|$ is the product of all binomials $x_i - x_j$ with $i > j$:

$$|V| = [(x_2 - x_1)] \cdot [(x_3 - x_2)(x_3 - x_1)] \cdots [(x_n - x_{n-1}) \cdots (x_n - x_1)].$$

3. Find the adjoint of a general 2×2 matrix A, and then compute the products $A(\text{adj } A)$ and $(\text{adj } A)A$.

4. Compute the adjoints of the matrices in Exercise 3 of Section 4–2.

5. If (x_1, y_1) and (x_2, y_2) are distinct points of the real plane, show that the equation

$$\begin{vmatrix} x & y & 1 \\ x_1 & y_1 & 1 \\ x_2 & y_2 & 1 \end{vmatrix} = 0$$

represents a straight line through these two points.

4–7 Nonsingularity and products. We have discovered the effect on $|A|$ of performing an elementary operation of type I or II on A. Now we are in position to study the third type of elementary operation.

THEOREM 4–7. *If B is obtained from A by an elementary row or column operation of type* III, $|B| = |A|$.

If $B = E_{hj}(c)A$, the rows of B are those of A except for row h, which is

$$a_{h1} + ca_{j1}, \ldots, a_{hn} + ca_{jn}.$$

If $|B|$ is expanded by the elements of row h, the cofactors c_{hk} employed are the same as the cofactors c_{hk} of the elements a_{hk} in A. Then Theorem 4–6 gives

$$|B| = \sum_k (a_{hk} + ca_{jk})c_{hk}$$
$$= \sum_k a_{hk}c_{hk} + c\sum_k a_{jk}c_{hk}$$
$$= |A| + c \cdot 0 = |A|.$$

The result for column operations may be proved by use of transposes.

The three results on elementary row operations may be summarized compactly. In each of the three cases we have $B = EA$, and $|EA|$ has the values

$$(23) \qquad -|A|, \quad c \cdot |A|, \quad |A|,$$

for types I, II, and III, respectively. Taking A to be the identity matrix we find from (23) that the determinant of an elementary matrix E is -1, $c \neq 0$, or 1 according as E is of type I, II, or III. Then in every case (23) yields the formula, $|EA| = |E| \cdot |A|$.

LEMMA 4–3. *If E is an elementary matrix and A is any square matrix of the same size, $|EA| = |E| \cdot |A|$.*

For a product $E_1 \cdots E_t B$ with elementary matrices E_j, repeated application of this lemma gives

$$(24) \qquad |E_1 \cdots E_t B| = |E_1| \cdots |E_t| \cdot |B|.$$

It is worth remembering that Theorems 4–1, 4–2, and 4–7 are all included in Lemma 4–3. We need only use the facts that $|E| = -1$, c, or 1 in the three cases, respectively.

THEOREM 4–8. *A square matrix A is nonsingular if and only if $|A| \neq 0$. In this case*

$$A^{-1} = \frac{1}{|A|} \cdot \text{adj } A.$$

If $|A| \neq 0$, Theorem 4–6 gives rise to the formula above for A^{-1}, so that A is nonsingular. Conversely, if A is nonsingular it is a product $A = E_1 \cdots E_t$ of elementary matrices E_j. Then (24) with $B = I$ gives

$$|A| = |E_1| \cdots |E_t|,$$

and this product is not zero, since each factor $|E_j|$ has value ± 1 or $c_j \neq 0$.

THEOREM 4–9. $|AB| = |A| \cdot |B|$.

If A is singular, $|A| = 0$ and the right side of the formula above is zero.   But the rank of AB is no greater than that of A (Exercise 7, Section 3–9) so that AB is singular and $|AB| = 0$.   This leaves the case in which A is nonsingular.   Then A is a product $E_1 \cdots E_t$ of elementary matrices E_j, and by (24),

$$|AB| = |E_1 \cdots E_t B| = |E_1| \cdots |E_t| \cdot |B| = |A| \cdot |B|.$$

EXERCISES

1. Show without use of Theorem 4–9 that $|AE| = |A| \cdot |E|$ if E is any elementary matrix.

2. Use Theorem 4–7 to calculate the determinants of the following matrices:

$$\begin{bmatrix} 1 & 3 & 5 \\ 2 & 4 & 4 \\ 1 & -1 & 1 \end{bmatrix}; \quad \begin{bmatrix} 5 & -6 & -1 & 1 \\ 0 & 2 & -3 & 2 \\ 1 & 2 & -1 & 4 \\ -1 & 0 & 2 & 1 \end{bmatrix}.$$

3. Calculate the inverse of the first matrix in Exercise 2 by Theorem 4–8 and also by the method of Theorem 3–16.

★4. If A is nonsingular, show that $|A^{-1}| = |A|^{-1}$.

5. Prove that the adjoint of a singular matrix is singular.

6. Prove that $|\text{adj } A| = |A|^{n-1}$ if A is $n \times n$.

7. Prove that $|\text{adj (adj } A)| = |A|^m$, $m = (n-1)^2$, where A is $n \times n$.

★8. Vandermonde's matrix (Exercise 2, Section 4–6) may be defined as the $n \times n$ matrix $V = (v_{ij})$, $v_{ij} = x_i^{j-1}$.   Prove that V is nonsingular if and only if $x_1, \ldots, x_n$ are distinct.

★9. Let $c_1, \ldots, c_n$ be distinct quantities of $\mathfrak{F}$, and let $k_1, \ldots, k_n$ be any quantities of $\mathfrak{F}$.   Prove that there is a polynomial $f(x) = a_0 + a_1x + \cdots + a_{n-1}x^{n-1} + x^n$ such that $f(c_i) = k_i$, $i = 1, \ldots, n$, and all the coefficients a_i lie in $\mathfrak{F}$.   (Hint: Use Exercise 8.)   Is $f(x)$ unique?

4–8 Determinants and rank.   The rank of any rectangular matrix may be characterized in terms of determinants of square submatrices, and this is, in fact, the way most of the older works treat rank.   The property is given here as the next theorem.

THEOREM 4–10.   *If A is any rectangular matrix, let t be the largest integer such that A has a $t \times t$ submatrix with determinant not zero. Then t is the rank of A.*

Let C be a $t \times t$ submatrix of A with $|C| \neq 0$.   Then C is nonsingular by Theorem 4–8, so that its rows, $\gamma_1, \ldots, \gamma_t$ are linearly independent. These rows are parts of possibly longer rows, $\alpha_1, \ldots, \alpha_t$, of A. If a linear combination vanishes,

$$c_1\alpha_1 + \cdots + c_t\alpha_t = 0,$$

then surely $c_1\gamma_1 + \cdots + c_t\gamma_t = 0$, so that all the c_j are zero. This shows that $\alpha_1, \ldots, \alpha_t$ are linearly independent, whence A has rank $r \geq t$.

Since A has rank r, it has r linearly independent rows forming an $r \times s$ submatrix B of rank r. Then B must have r linearly independent columns forming an $r \times r$ submatrix C of B and of A. Since the columns of C are independent, C has rank r and is nonsingular, so that $|C| \neq 0$. This proves $t \geq r$. With these two inequalities we conclude that $r = t$.

To compute the rank of A by Theorem 4–10 we would have to find a $t \times t$ submatrix which is nonsingular, then show that every larger square submatrix is singular. It would suffice (why?) to show that every $(t + 1) \times (t + 1)$ submatrix is singular, but the number of submatrices to examine can be reduced still further, as the next result shows.

THEOREM 4–11. *Let A have a $t \times t$ submatrix T which is nonsingular. Let every submatrix of A containing T and one more row and one more column be singular. Then A has rank t.*

It will simplify the notation without destroying the generality of the argument if we take T in the first t rows and columns of $A = (a_{ij})$:

$$A = \begin{bmatrix} a_{11} & \ldots & a_{1s} \\ \cdot & & \cdot \\ \cdot & & \cdot \\ \cdot & & \cdot \\ a_{n1} & \ldots & a_{ns} \end{bmatrix}, \quad T = \begin{bmatrix} a_{11} & \ldots & a_{1t} \\ \cdot & & \cdot \\ \cdot & & \cdot \\ \cdot & & \cdot \\ a_{t1} & \ldots & a_{tt} \end{bmatrix}.$$

Let α_i designate row i in A, and τ_i designate that part of α_i belonging to the first t columns:

$$\alpha_i = (a_{i1}, \ldots, a_{is}), \quad \tau_i = (a_{i1}, \ldots, a_{it}),$$

where $i = 1, \ldots, n$ in each case. The rows $\tau_1, \ldots, \tau_t$ of T are linearly independent and span $V_t(F)$. If $q > t$, τ_q is then uniquely expressible as a linear combination of $\tau_1, \ldots, \tau_t$:

$$\tau_q = c_1\tau_1 + \cdots + c_t\tau_t.$$

Now suppose that U is a submatrix of A constructed from the submatrix T by use also of row q and column r of A, $q > t, r > t$:

$$U = \begin{bmatrix} \tau_1, & a_{1r} \\ \tau_2, & a_{2r} \\ \cdot & \cdot \\ \cdot & \cdot \\ \cdot & \cdot \\ \tau_t, & a_{tr} \\ \tau_q, & a_{qr} \end{bmatrix}.$$

By hypothesis, U is singular, its rows are linearly dependent, and it must be possible to express its last row (why?) as a linear combination of the preceding rows:

$$(\tau_q, a_{qr}) = k_1(\tau_1, a_{1r}) + \cdots + k_t(\tau_t, a_{tr}).$$

This implies that

$$\tau_q = k_1\tau_1 + \cdots + k_t\tau_t, \quad a_{qr} = k_1a_{1r} + \cdots + k_ta_{tr}.$$

By the uniqueness of $c_1, \ldots, c_t$ above, we conclude that $k_1 = c_1, \ldots, k_t = c_t$. The coefficients k_i above are thus independent of r, that is, are the same set of t scalars for all choices of $r = t+1, \ldots, s$. Therefore, if we recall that

$$\alpha_q = (\tau_q, a_{q,t+1}, \ldots, a_{qs}),$$

we conclude that

$$\alpha_q = c_1\alpha_1 + \cdots + c_t\alpha_t.$$

Since the rows of A are thus linear combinations of t of these rows, A has rank $\leq t$. By the preceding theorem the rank is $\geq t$, whence it is exactly t.

Exercises

1. Compute by determinants the ranks of the matrices in Exercise 2, Section 4–7.

2. Prove by determinants that the vector space in $V_4(\Re)$ spanned by

$$\xi = (0, 1, 2, 3), \quad \eta = (1, 1, 2, 2),$$

has dimension two.

3. Use Theorem 4–10 to prove that the rank and nullity of a matrix do not change when the scalar field is enlarged.

4. Answer the question raised in the proof of Theorem 4–11.

4–9 Cramer's rule. The adjoint formula for A^{-1} in Theorem 4–8 has a consequence for systems

$$(25) \qquad AX = K, \quad K = \text{col } (k_1, \ldots, k_n)$$

of n linear equations in n unknowns $x_1, \ldots, x_n$ with nonsingular co-efficient matrix $A = (a_{hj})$. Multiplying on the left by A^{-1} gives the solution in the form

$$(26) \qquad X = A^{-1}K$$

and shows that the solution is unique. By Theorem 4–8 this may be written as

$$(27) \qquad X = \frac{1}{|A|} (\text{adj } A)K.$$

Row h on the left side of (27) is simply the unknown x_h. Since row h of adj A consists of the cofactors $c_{1h}, \ldots, c_{nh}$, (27) yields the following formula:

$$(28) \qquad x_h = \frac{c_{1h}k_1 + \cdots + c_{nh}k_n}{|A|}. \qquad (h = 1, \ldots, n)$$

Consider the matrix $A(h)$ constructed from A by replacing column h by the column, K, of constant terms. The cofactor of k_j in $A(h)$ is the same as the cofactor c_{jh} of a_{jh} in A, so that the numerator on the right side of (28) is the expansion of $A(h)$ by the elements of column h. Thus (28) becomes

$$(29) \qquad x_h = \frac{|A(h)|}{|A|}. \qquad (h = 1, \ldots, n)$$

Formula (29), or the equivalent (28), for solving a system (25) of linear equations with nonsingular coefficient matrix is known as Cramer's rule. Its main claim to fame is that it gives a routine procedure for finding each unknown, and that its application requires very little mathematical background. Neither (29) nor (26) is a very efficient rule for numerical solutions when n is large, unless high-speed computing machines are used.

EXERCISES

1. By Cramer's rule, solve the system

$$\begin{aligned} x + y + z &= 0, \\ 2x - y + z &= -3, \\ x + 2y &= 5. \end{aligned}$$

2. Solve the system in Exercise 1 by the method indicated in (26).

CHAPTER 5

CONGRUENCE AND HERMITIAN CONGRUENCE

5–1 Quadratic forms. A polynomial like

(1) $$2x^2 + 6xy - 7y^2 = f(x, y) = f,$$

with real coefficients, every term being of degree two in x and y, is called a quadratic form in x and y. To relate (1) to matrices, we write it as

$$
\begin{aligned}
f &= 2x^2 + 3xy + 3yx - 7y^2 \\
&= x(2x + 3y) + y(3x - 7y) \\
&= (x, y)\begin{bmatrix} 2x + 3y \\ 3x - 7y \end{bmatrix} = (x, y)\left[\begin{pmatrix} 2 & 3 \\ 3 & -7 \end{pmatrix}\begin{pmatrix} x \\ y \end{pmatrix}\right].
\end{aligned}
$$

Thus the matrix

(2) $$A = \begin{bmatrix} 2 & 3 \\ 3 & -7 \end{bmatrix}$$

has been associated with the form (1), and f may be regarded as a 1×1 matrix which is the product,

(3) $$f = X'AX, \quad X = \text{col } (x, y).$$

Notice that the diagonal elements, 2 and -7, in A are the coefficients of the squared terms in f, and the other elements, both equal to 3, are the coefficients of the two equal terms

$$3xy + 3yx,$$

into which the cross-product term of f has been split. Notice, also, that A has the property

(4) $$A' = A.$$

A square matrix $A = (a_{ij})$ over any field $\mathfrak{F}$ is called *symmetric* if, as in (4), it is equal to its transpose. An equivalent property is that $a_{ij} = a_{ji}$ for all i and j.

A polynomial $f = f(x_1, \ldots, x_n)$ in variables $x_1, \ldots, x_n$ with coefficients in a field $\mathfrak{F}$ is called a *quadratic form* in $x_1, \ldots, x_n$ if every term of f is quadratic in the variables. Then f may be written as

(5) $$f = \Sigma x_i a_{ij} x_j. \qquad (i, j = 1, \ldots, n)$$

84

This notation assumes that each term in $x_h x_k$ ($h \neq k$) has been split into two terms

$$a_{hk} x_h x_k + a_{kh} x_k x_h.$$

There are many ways of doing this. The split-up which we wish to use is that in which $a_{hk} = a_{kh}$, so that each of these equal quantities is half of the total coefficient c_{hk} of $x_h x_k$ in f. With this agreement, every quadratic form f in $x_1, \ldots, x_n$ is associated with a unique matrix A such that

$$(6) \qquad A' = A, \quad f = X'AX, \quad X = \text{col}(x_1, \ldots, x_n).$$

This matrix A is called the *coefficient matrix* of f, or simply the *matrix* of f.

Every field contains a quantity 1, and 2 is defined to be $1 + 1$. The elementary algebra of finding the coefficients discussed above proceeds as follows:

$$a_{hk} = a_{kh}, \quad a_{hk} + a_{kh} = c_{hk},$$
$$2a_{hk} = c_{hk}, \qquad a_{kh} = \tfrac{1}{2} c_{hk}.$$

In the last step we multiply by $\tfrac{1}{2}$, that is, the inverse of the quantity 2. This is perfectly valid in any subfield of $\mathcal{C}$, and in most other fields. There are, however, certain unusual fields in which $1 + 1 = 0$. For these fields 2 is 0, hence has no inverse, and the process above is invalid. In our discussion of quadratic forms and symmetric matrices we shall always make the assumption on $\mathcal{F}$ that $1 + 1 \neq 0$. (The reader who is interested only in the cases of $\mathcal{R}$ and $\mathcal{C}$ may disregard this hypothesis.) To repeat, then, every quadratic form over a field $\mathcal{F}$ in which $1 + 1 \neq 0$, has a unique, symmetric coefficient matrix.

Quadratic forms occur not only in the study of conic sections and quadric surfaces in analytic geometry, but also in problems of maxima and minima, dynamics, and statistics, as well as throughout higher mathematics. This is perhaps the reason that the theories of quadratic forms and symmetric matrices over $\mathcal{R}$ and $\mathcal{C}$ were highly developed at an early stage.

If the form f in (1) is placed equal to a scalar, say to 1, the equation of a conic is obtained:

$$2x^2 + 6xy - 7y^2 = 1.$$

The routine procedure for determining the nature of this conic re-

quires a rotation of axes to eliminate the cross-product term. This
is done by a substitution:

$$x = x' \cos t + y' \sin t$$
$$y = -x' \sin t + y' \cos t$$

or

$$\begin{bmatrix} x \\ y \end{bmatrix} = \begin{bmatrix} \cos t & \sin t \\ -\sin t & \cos t \end{bmatrix} \begin{bmatrix} x' \\ y' \end{bmatrix}.$$

When t is appropriately chosen, the conic is represented by a new
equation in x' and y' with only squared terms present.

Wherever quadratic forms (6) arise, one is usually interested in
"simplifying" them by making a nonsingular linear substitution

(7) $$X = PY,$$

where $P = (p_{ij})$ is a nonsingular matrix and Y is an n-tuple of new
variables, $Y = \text{col } (y_1, \ldots, y_n)$. Such linear substitutions always
lead to quadratic forms in $(y_1, \ldots, y_n)$. In fact (7) may be substituted
directly in (6):

$$f = X'AX = (PY)'APY = Y'P'APY.$$

Notice that $P'AP$ is symmetric (A is symmetric by hypothesis):

(8) $$(P'AP)' = P'A'(P')' = P'AP.$$

Since the symmetric coefficient matrix of a quadratic form is unique,
we have:

THEOREM 5–1. *If a quadratic form with matrix A is subjected to a
nonsingular linear substitution* (7), *it becomes a quadratic form with
matrix $P'AP$, P nonsingular.*

Two matrices B and A are called *congruent* if

(9) $$B = P'AP$$

for some nonsingular P. The reduction (or simplification) of quadratic forms by nonsingular linear substitutions amounts to the same
thing as the reduction of symmetric matrices by congruence. Observe that B and A are necessarily square and of the same size as P.
Moreover, congruent matrices are equivalent, hence have the same
rank.

EXERCISES

1. Write the matrix of each of the following forms:
   (a) $x_1^2 + 2x_1x_2 + 3x_1x_3 + 4x_2^2 + 5x_3^2$;
   (b) $2xy$.

★2. Show that congruence of $n \times n$ matrices is an RST relation.

3. Show that $A'A$ and AA' are always symmetric, A being any rectangular matrix.

4. Let A and B be symmetric matrices of the same size. Show that AB is symmetric if and only if A commutes with B.

5. Show that every nonsingular symmetric matrix is congruent to its inverse.

6. If A is symmetric, show that $B = c_m A^m + \cdots + c_1 A + c_0 I$ is symmetric, the c_i being arbitrary scalars. Show also that A^{-1} is symmetric if it exists.

7. Let A and B be congruent as in (9), where P is nonsingular. Prove that all nonsingular matrices Q obeying $Q'AQ = B$ are given by $Q = RP$, R varying over all nonsingular matrices such that $R'AR = A$. (Such matrices R are called *congruent automorphs* of A.)

8. Let $f = a_1 x_1 + \cdots + a_n x_n$. Show that the quadratic form f^2 in $x_1, \ldots, x_n$ has coefficient matrix $C = \xi'\xi$, where ξ is a suitable row vector. Show also that $C = A'A$, where A is the $n \times n$ matrix with ξ as its first row, all other rows being zero.

9. Let $g = f_1^2 + \cdots + f_r^2$, where $f_k = a_{k1} x_1 + \cdots + a_{kn} x_n$ and $r \leq n$. Show that $g = X'CX$, $X = \text{col}(x_1, \ldots, x_n)$, where $C = A'A$ and A is an $n \times n$ matrix generalizing the matrix A of Exercise 8. Then show that $|C| = 0$ if $r < n$.

5-2 Range of values. If $f = f(x_1, \ldots, x_n)$ is a quadratic form over a field $\mathfrak{F}$ and if $a_1, \ldots, a_n$ are in $\mathfrak{F}$, $f(a_1, \ldots, a_n)$ is a *value* of f. The totality (or range) of values of f is some subset of $\mathfrak{F}$.

THEOREM 5-2. *The range of values of a quadratic form does not change under nonsingular linear substitution.*

We have

$$f = X'AX = Y'P'APY, \quad X = PY,$$

where P is nonsingular. Notice that "old" and "new" forms may both be regarded as f, but expressed in terms of different variables which are related by the equation $X = PY$. If a value of f is obtained by replacing Y by the vector Y_0 of values in $\mathfrak{F}$, the same value of f is obtained from the old variables by replacing X by $X_0 = PY_0$. Conversely, each value obtained by using $X = X_0$ is also obtainable by using $Y = Y_0 = P^{-1}X_0$. This proves the theorem.

To say that two quadratic forms f and g in the same variables $x_1, \ldots, x_n$ are equal is to say that they have the same coefficient matrices. Then they surely have the same range of values and, moreover,

(10) $$f(a_1, \ldots, a_n) = g(a_1, \ldots, a_n)$$

for all a_i in $\mathfrak{F}$, $i = 1, \ldots, n$. Now we answer a converse question.

THEOREM 5–3. *Let $\mathfrak{F}$ be a field in which $1 + 1 \neq 0$. Let f and g be quadratic forms in $x_1, \ldots, x_n$ over $\mathfrak{F}$ such that (10) holds for all a_i in $\mathfrak{F}$, $i = 1, \ldots, n$. Then f and g are the same form.*

If

$$f = X'AX, \quad g = X'BX,$$

it suffices to show that $A = B$. Also

$$h = f - g = X'(A - B)X$$

is a quadratic form whose range of values consists only of zero. Hence it suffices to show that if $h = X'CX$ has 0 as its range of values, then $C = 0$. Let $C = (c_{ij})$. If X is replaced by the unit vector u_i there results

$$0 = u_i'Cu_i = c_{ii}, \qquad (i = 1, \ldots, n)$$

so that the diagonal elements of C are all zero. Next replace X by $u_i + u_j$ $(i \neq j)$:

$$\begin{aligned} 0 &= (u_i' + u_j')C(u_i + u_j) \\ &= c_{ii} + c_{ji} + c_{ij} + c_{jj} \\ &= c_{ji} + c_{ij} = 2c_{ij}, \end{aligned}$$

since C is symmetric and h has only the value 0. Thus $2c_{ij} = 0$, $c_{ij} = 0 = c_{ji}$, and $C = 0$.

DEFINITION 1. A quadratic form f over $\mathfrak{R}$ is called *positive semidefinite* if none of its values is negative; it is *positive definite* if its values are positive except when all of the variables are zero.

The form $x^2 + y^2$ in variables x and y over $\mathfrak{R}$ is clearly positive definite. The real form

$$f = x^2 - xy + \tfrac{1}{4}y^2$$

is 0 for $x = 1$, $y = 2$, so that it cannot be positive definite. However

$$f = (x - \tfrac{1}{2}y)^2$$

so that f is never negative; f is positive semidefinite.

DEFINITION 2. A symmetric matrix A over $\mathfrak{R}$ is *positive semidefinite* or *positive definite* if the form defined by A has the like-named property.

The nature of these matrices will be scrutinized closely in Section 5–5.

1. If the scalar field is $\Re$, find the range of values of each of the following forms in x and y:

(a) $x^2 + y^2$;   (b) $x^2 - y^2$;   (c) xy.

What are their ranges if $\mathfrak{C}$ is the field of scalars?

2. Show that the form $x^2 + xy$ over $\Re$ is not positive semidefinite.

3. Show that matrix $\begin{bmatrix} 1 & -1 \\ -1 & 1 \end{bmatrix}$ is positive semidefinite.

4. If two real, quadratic forms in $x_1, \ldots, x_n$ are positive definite, show that their sum is positive definite.

5. If f is a positive definite quadratic form in $x_1, \ldots, x_r$, and g is a positive definite quadratic form in $x_{r+1}, \ldots, x_n$, show that $f + g$ is a positive definite quadratic form in $x_1, \ldots, x_n$.

5–3 Congruent symmetric matrices. Congruence is an RST relation on the set of all $n \times n$ matrices over $\mathfrak{F}$. Most of the known results about the matrices congruent to a given matrix A are confined to the case in which A is symmetric. In this case, as noted in (8), every matrix congruent to A is also symmetric. With this as a start we proceed to seek out the simplest possible matrices congruent to a given matrix.

THEOREM 5–4. *Two matrices are congruent if and only if one is obtainable from the other by a succession of pairs of elementary operations, each pair consisting of a column operation and the corresponding row operation. In each pair either operation may be performed first.*

Each nonsingular matrix P is a product of elementary matrices:

$$P = E_1 \cdots E_k.$$

Hence $B = P'AP$ if and only if

$$B = E_k'[\cdots (E_1'AE_1) \cdots]E_k.$$

For each column operation effected by E_i as a right factor the corresponding row operation is effected by E_i' as a left factor. The final statement in the theorem merely amounts to the associative law:

$$E_i'(CE_i) = (E_i'C)E_i.$$

THEOREM 5–5. *Over any field $\mathfrak{F}$ in which $1 + 1 \neq 0$, each symmetric matrix A is congruent to a diagonal matrix in which the number of nonzero diagonal elements is the rank of A.*

If $A = 0$, the desired diagonal matrix is A itself. Hence we assume $A \neq 0$. Next we show that A is congruent to a matrix B with a nonzero diagonal element. This is true with $B = A = I'AI$ if $A = (a_{hj})$ has a nonzero diagonal element a_{hh}. Now we may assume that every $a_{hh} = 0$, whence some $a_{hj} \neq 0$ $(h \neq j)$. Then

$$a_{hj} = a_{jh} \neq 0, \quad a_{hh} = a_{jj} = 0.$$

Addition of column j of A to column h, then of row j to row h, replaces A by a congruent matrix $B = (b_{kj})$, in which

(11) $b_{hh} = a_{hj} + a_{jh} = 2a_{hj} \neq 0.$

Thus in any case there is a matrix B congruent to A with a nonzero diagonal element b_{hh}. Interchange of column h and column 1 in B followed by the corresponding row operation replaces B by $C = (c_{kj})$ with $c_{11} = b_{hh} \neq 0$. Now we add to column j the product of the first column by $-c_{1j}/c_{11}$, then perform the corresponding row operation. Since $c_{j1} = c_{1j}$, this makes the new elements in the $(1, j)$ and $(j, 1)$ places both 0. Doing this for each j, we replace C by a matrix

$$D = \begin{bmatrix} c_{11} & 0 \\ 0 & D_1 \end{bmatrix},$$

where D_1 has $n - 1$ rows and columns and is symmetric. (Is computation necessary to verify the symmetry of D_1?)

The same procedure may be applied to D_1 by applying it to D with operations not affecting the first row or column. After a finite number of steps there appears a diagonal matrix congruent to A. Preservation of rank is a consequence of the fact that congruence is a special case of equivalence.

EXERCISES

1. Using elementary operations, transform the following real, symmetric matrices to congruent matrices with at least one nonzero diagonal element:

$$A_1 = \begin{bmatrix} 0 & 1 & 2 \\ 1 & 0 & 3 \\ 2 & 3 & 0 \end{bmatrix}; \quad A_2 = \begin{bmatrix} 0 & 0 & 1 \\ 0 & 0 & 0 \\ 1 & 0 & 0 \end{bmatrix}.$$

Then complete the reduction of each of these matrices to a congruent diagonal matrix. Express the results in terms of quadratic forms f_1 and f_2.

2. Find a nonsingular matrix P with rational elements such that $P'AP$ is diagonal, where

$$A = \begin{bmatrix} 1 & 2 \\ 2 & 4 \end{bmatrix}.$$

3. Prove that a symmetric, complex matrix of rank r is congruent to

$$\begin{bmatrix} I_r & 0 \\ 0 & 0 \end{bmatrix}.$$

Hence prove that two $n \times n$ complex, symmetric matrices are congruent if and only if they have the same rank.

5–4 Congruence over ℜ. The diagonal matrix of Theorem 5–5 is not uniquely determined by the given matrix A. In fact the three matrices

$$A = \begin{bmatrix} 1 & 0 \\ 0 & 1 \end{bmatrix}, \quad B = \begin{bmatrix} 4 & 0 \\ 0 & 4 \end{bmatrix}, \quad C = \begin{bmatrix} 9 & 0 \\ 0 & 9 \end{bmatrix}$$

are congruent over the rational number field. (Write matrices P effecting these congruences!) Over the field of reals it is possible to strengthen the result of the last theorem.

Theorem 5–6. *A real symmetric matrix A of rank r is congruent* to a matrix*

$$B = \begin{bmatrix} I_p & 0 & 0 \\ 0 & -I_{r-p} & 0 \\ 0 & 0 & 0 \end{bmatrix}.$$

The integer p is uniquely determined by A.

By the last theorem A is congruent to a matrix of the type

$$D = \text{diag } (d_1, \ldots, d_r, \ 0, \ldots, 0),$$

where each d_i is nonzero. Let p designate the number of d_h which are positive. By suitable interchanges of rows followed by the corresponding column interchanges, these p elements may be brought into the first p positions. Since the resulting matrix is congruent to D, assume that D itself has the properties, $d_h > 0$ $(h = 1, \ldots, p)$ and $d_j < 0$ $(j > p)$. Then there are real numbers c_k such that

$$c_h^2 = d_h^{-1}, \quad c_j^2 = -d_j^{-1}, \qquad \begin{bmatrix} h = 1, \ldots, p \\ j = p + 1, \ldots, r \end{bmatrix}$$
$$c_{r+1} = \cdots = c_n = 1.$$

---

* In any such statement it will be understood that the matrix P effecting the congruence can be chosen to have elements in ℜ. This convention will permit us to avoid tedious repetition of the phrase "over ℜ."

The matrix $P = \text{diag} (c_1, \ldots, c_n)$ is nonsingular and $P'DP$ is the matrix B of the theorem.

To prove the uniqueness, suppose that A is also congruent to

$$C = \begin{bmatrix} I_q & 0 & 0 \\ 0 & -I_{r-q} & 0 \\ 0 & 0 & 0 \end{bmatrix}.$$

Then the form $f = X'AX$ is convertible by nonsingular linear substitutions

(12) $$X = PY, \quad X = QZ,$$

into

(13) $$y_1^2 + \cdots + y_p^2 - y_{p+1}^2 - \cdots - y_r^2$$

and

(14) $$z_1^2 + \cdots + z_q^2 - z_{q+1}^2 - \cdots - z_r^2.$$

If $p \neq q$, it is only a matter of notation to assume $q < p$.

If we express the new variables in terms of the old,

(15) $$Y = P^{-1}X, \quad Z = Q^{-1}X,$$

each y_i and each z_h is a unique linear combination of $x_1, \ldots, x_n$. The equations

(16) $$\begin{aligned} z_h &= 0, & (h = 1, \ldots, q) \\ y_j &= 0 & (j = p + 1, \ldots, n) \end{aligned}$$

may then be regarded as simultaneous linear, homogeneous equations in $x_1, \ldots, x_n$. Being $q + n - p$ in number, $q + n - p < n$, these homogeneous equations have a nontrivial solution (Corollary 3–6):

(17) $$X_0 = \text{col} (x_1, \ldots, x_n) \neq (0, \ldots, 0).$$

With $X = X_0$ the value of f is $X_0'AX_0$, but this value may also be computed from (13) by use of $Y_0 = P^{-1}X_0$, or from (14) by use of $Z_0 = Q^{-1}X_0$. From (13) we see that $f \geq 0$, since (16) requires those y_i with minus signs to be 0. Likewise (14) and (16) show that $f \leq 0$. Hence $f = 0$. But by (16)

$$f = y_1^2 + \cdots + y_p^2 = 0,$$

so that $y_1 = \cdots = y_p = 0$ and $y_{p+1} = \cdots = y_n = 0$. Then $Y_0 = 0$, $X_0 = PY_0 = 0$ in conflict with (17). This contradiction, arising from the assumption that $p \neq q$, completes the proof.

DEFINITION 3.   The integer p of Theorem 5–6 is called the *index* of the symmetric, real matrix A.

If A_1 is congruent to A of Theorem 5–6, it is also congruent to B (transitivity of congruence!) so that A_1 has the same rank r and index p as A. Conversely, if A_1 has the same size, rank, and index as A, they are congruent to the same matrix B in Theorem 5–6, and hence to each other. Thus

THEOREM 5–7. *Two $n \times n$ real, symmetric matrices are congruent if and only if they have the same rank and the same index.*

Since congruence of real, symmetric matrices is an RST relation, it separates this class into subclasses, each of which consists of all such matrices which are congruent to one matrix. The last theorem' shows that each class is distinguished by three integers, n, r, and p: all members in one subclass have the same number n of rows, the same rank r, and the same index p; no two subclasses have the same values for all three of these integers. Thus n, r, and p are called a *complete set of invariants* for real, symmetric matrices relative to congruence. The matrices in Theorem 5–6 provide a canonical set.

EXERCISES

1. What are the ranks and indices of the matrices A_1 and A_2 in Exercise 1, Section 5–3?   Of the matrices of the quadratic forms $x_1 x_2$ and $x_1 x_2 + x_3 x_4$?

2. On the assumption that $\mathfrak{R}$ is the scalar field, interpret Exercise 5, Section 5–1, in terms of rank and index.

3. Let A be an $n \times n$ real symmetric matrix of index p.   Prove that $|A|$ is positive if and only if A is nonsingular and $n - p$ is even.

4. Prove the first part of Theorem 5–6 by working directly with substitutions in a quadratic form, starting with the result of Theorem 5–5.

5. List all 4×4 real, symmetric matrices which are in the canonical form of Theorem 5–6.

6. Show that the number of $n \times n$ real, symmetric matrices which are in the canonical form of Theorem 5–6 is $\frac{1}{2}(n + 1)(n + 2)$.

5–5 Positive semidefinite matrices. Since the range of values of a quadratic form does not change under nonsingular linear substitution, a form (or real symmetric matrix) is positive semidefinite if and only if its canonical form (or matrix) has this property.   The form f associated with B in Theorem 5–6 is displayed in (13), and f is positive semidefinite if and only if $p = r$.   If $p = r < n$, (13) becomes

$$f = y_1^2 + \cdots + y_p^2 + 0y_{p+1}^2 + \cdots + 0y_n^2,$$

so that $f = 0$ with $y_1 = \cdots = y_{n-1} = 0$, $y_n = 1$.   Hence if f is positive definite, $p = r = n$.   The converse is apparent.   This gives

THEOREM 5–8. *An $n \times n$ real, symmetric matrix of rank r and index p is positive semidefinite if and only if $p = r$, and positive definite if and only if $p = r = n$.*

In other words, A is positive definite if and only if its canonical matrix of Theorem 5–6 is I. Then A is congruent to I,

$$(18) \qquad\qquad A = P'IP = P'P,$$

where P is real and nonsingular. Conversely, (18) implies that A has rank and index n. (Why?) This gives

THEOREM 5–9. *A real matrix A is positive definite if and only if there is a nonsingular real matrix P such that $A = P'P$.*

A submatrix of a square matrix A is called *principal* if it is obtained by deleting certain rows and the like-numbered columns. The determinant of a principal submatrix is called a *principal subdeterminant* or principal minor.

THEOREM 5–10. *If A is positive definite, every principal submatrix is positive definite. Also $|A|$ and all principal subdeterminants are positive.*

If $f = X'AX$, the principal submatrix S obtained by deleting row h and column h from A is the matrix of the form g obtained from f by putting $x_h = 0$. Then every value of g for values of its variables which are not all zero is also such a value of f, hence is positive by the hypothesis on f and A. This proves g and S positive definite.

The procedure just applied to f and A may be applied to g and S, with the result that every principal submatrix of A obtained by deleting two rows and columns is positive definite. A repetition shows that the positive definite property of A carries over to all of the principal submatrices of A.

By Theorems 5–9 and 4–9

$$|A| = |P'P| = |P'| \cdot |P| = |P|^2 > 0.$$

Thus every positive definite matrix has positive determinant. This fact together with the first part of Theorem 5–10 proves the last part.

EXERCISES

1. A quadratic form f and its matrix A are called *negative semidefinite* if all values of f are $\leqq 0$, and *negative definite* if these values are negative except when all the variables are zero. State and prove the analogue of Theorem 5–8 for these concepts.

2. Let f be a quadratic form over $\Re$. If f assumes both positive and negative values, show that f can be given the value zero for values of the variables which are not all zero. State and prove a partial converse.

3. Let f be a quadratic form over $\Re$ with nonsingular coefficient matrix. Prove that f is positive or negative definite if and only if $f \neq 0$ whenever the variables are given values which are not all zero.

4. Prove that the real quadratic form $ax_1^2 + 2bx_1x_2 + cx_2^2$ is positive definite if and only if $a > 0$ and $|A| > 0$, where A is the matrix of the quadratic form.

5. An expression $c_1x_1 + \cdots + c_nx_n$ is a *linear form* in $x_1, \ldots, x_n$, and is a *nonzero* linear form if at least one coefficient c_i is different from zero. Prove that if a quadratic form $f = X'AX$ is the square of a nonzero linear form, A has rank one. Is the converse true without restrictions on the field of scalars?

6. In Exercise 5 let $\Re$ be the field of scalars. Prove that f is a product of two nonzero linear forms if and only if A has rank one, or rank two and index one.

7. In Exercise 5 let $\mathbb{C}$ be the scalar field. Prove that f is a product of two nonzero linear forms if and only if A has rank one or two.

5–6 Skew matrices. A square matrix A is called *skew* (also *skew-symmetric* or *alternate*) if $A' = -A$. If $A = (a_{ij})$, this is equivalent to the requirement that $a_{ij} = -a_{ji}$ for all i and j. Then in particular each diagonal element a_{ii} is equal to its negative $-a_{ii}$, so that $2a_{ii} = 0$ and $a_{ii} = 0$, provided that $1 + 1 \neq 0$. The latter assumption will always be made here, so that we can conclude that skew matrices have zero diagonals.

Each matrix $B = P'AP$ congruent to a skew matrix is also skew:

$$(P'AP)' = P'A'P = P'(-A)P = -(P'AP).$$

The class of skew matrices over a fixed field $\mathfrak{F}$ may thus be studied for canonical sets relative to congruence. In view of the fact that no such set is known for symmetric matrices except when the field is severely specialized, it is interesting and surprising to find that we can with very little effort get a canonical set of skew matrices over a general $\mathfrak{F}$ in which $1 + 1 \neq 0$.

THEOREM 5–11. *Each skew matrix A over a field $\mathfrak{F}$ (in which $1 + 1 \neq 0$)* is congruent over $\mathfrak{F}$ to a matrix $B = \text{diag} (E_1, \ldots, E_s, 0)$, where 0 is a zero matrix of suitable size (possibly absent),*

---

* The theorem and its proof are valid if this assumption on $\mathfrak{F}$ is replaced by a weaker assumption — namely, that the skew matrix A has zero diagonal (which is restrictive only when $1 + 1 = 0$).

$$E_i = \begin{bmatrix} 0 & 1 \\ -1 & 0 \end{bmatrix}, \qquad (i = 1, \ldots, s)$$

and $2s$ is the rank of A.

If $A = 0$, take $B = A$. Otherwise $A = (a_{ij}) \neq 0$ and every $a_{ii} = 0$, so that some

$$a_{ij} = -a_{ji} = c \neq 0.$$

Interchange of row i with row 1, and row j with row 2, followed by the corresponding column interchanges, replaces A by a congruent matrix C, also skew, such that

$$C = \begin{bmatrix} C_1 & C_2 \\ C_3 & C_4 \end{bmatrix}, \quad C_1 = \begin{bmatrix} 0 & c \\ -c & 0 \end{bmatrix}.$$

Multiplication of the first row and first column by c^{-1} may be effected by a congruence, so that we may assume that

$$C_1 = \begin{bmatrix} 0 & 1 \\ -1 & 0 \end{bmatrix} = E_1.$$

It is clear that the element -1 in C_1 may be exploited by row operations to make the first column of C_3 zero; the corresponding column operations make the first row of C_2 zero. (Do not compute to prove the latter statement. Merely use the fact that congruence replaces a skew matrix by a skew matrix.) Likewise the second column of C_3 may be made zero by use of the $+1$ in C_1, and the second row of C_2 may be made zero also. Thus A is congruent to a matrix

$$\begin{bmatrix} E_1 & 0 \\ 0 & A_1 \end{bmatrix}.$$

If $A_1 = 0$ the construction is completed. Otherwise the process may be repeated. This shows that A is congruent to a matrix B of the type described in the theorem. Since the first $2s$ rows of B are linearly independent, and the others zero, B has rank $2s$ and so does A. This completes the proof.

The reader may prove the following consequence of Theorem 5–11. It shows that rank and the number of rows form a complete set of invariants for skew matrices relative to congruence.

THEOREM 5–12. *Two $n \times n$ skew matrices over a field in which $1 + 1 \neq 0$ are congruent if and only if they have the same rank.*

There are many reasons for interest in skew matrices. One of them is that an arbitrary square matrix over $\mathfrak{F}$ (in which $1 + 1 \neq 0$) is expressible uniquely as

(19) $$A = S + K,$$

where S is symmetric and K is skew. Suitable choices of S and K are

$$S = \tfrac{1}{2}(A + A'), \quad K = \tfrac{1}{2}(A - A'),$$

as one finds by substituting these matrices in (19). Conversely, if (19) is any expression for A as the sum of a symmetric matrix S and a skew matrix K, it is easy to show that S and K must be the matrices above, thereby establishing the uniqueness of the decomposition (19).

EXERCISES

★1. Prove that the matrices S and K in (19) are uniquely determined by the given square matrix A.

2. Let E be the 2×2 matrix E_i employed in Theorem 5–11. Show that a 2×2 matrix P obeys $P'EP = E$ if and only if $|P| = 1$.

3. Find the decomposition (19) for each of the following matrices A:

$$\begin{bmatrix} 1 & 2 \\ 3 & 4 \end{bmatrix}; \quad \begin{bmatrix} 0 & 1 & 5 \\ -1 & 3 & 0 \\ 2 & 6 & 1 \end{bmatrix}.$$

4. Find a nonsingular matrix P with rational elements such that $P'AP$ is in the canonical form of Theorem 5–11, where

$$A = \begin{bmatrix} 0 & 1 & 2 \\ -1 & 0 & 3 \\ -2 & -3 & 0 \end{bmatrix}; \quad A = \begin{bmatrix} 0 & 1 & 1 & 1 \\ -1 & 0 & 1 & 0 \\ -1 & -1 & 0 & -2 \\ -1 & 0 & 2 & 0 \end{bmatrix}.$$

5. Prove that A^2 is symmetric if A is skew.

6. Let A and B be skew matrices of the same size. Show that AB is symmetric if and only if A commutes with B.

7. Prove that a skew matrix with an odd number of rows is singular.

8. Prove that the determinant of a real skew matrix is zero or positive.

5–7 Hermitian matrices. There is a type of complex matrix which generalizes the real symmetric matrices: $A = (a_{hj})$ is called *Hermitian* if it is square and, for all h and j,

(20) $$a_{jh} = \bar{a}_{hj},$$

where a symbol $\bar{a}$ always denotes the conjugate of the complex number a. Let

$$\bar{A} = (\bar{a}_{hj})$$

denote the matrix obtained from A by replacing each element by its conjugate. As may be expected, $\bar{A}$ is called the *conjugate* of A.

Then the definition of a Hermitian matrix is that it is a complex matrix A which is equal to its conjugate transpose:

$$\bar{A}' = A.$$

If A is real and Hermitian, it is symmetric.

Some facts about complex numbers $z = a + bi$ should be recalled. The conjugate of z is $\bar{z} = a - bi$, whence $z = \bar{z}$ if and only if z is real. Then (20) with $j = h$ shows that every diagonal element of a Hermitian matrix is real. Recall also that the sum and the product of complex conjugates are real:

$$z + \bar{z} = 2a, \quad z\bar{z} = a^2 + b^2.$$

Finally, we must know that the conjugate of a product is the product of the conjugates:

$$\overline{z_1 z_2} = \bar{z}_1 \cdot \bar{z}_2.$$

Hermitian matrices are of interest in connection with *Hermitian forms*,

$$f = \sum_{h,j}^{1,\ldots,n} \bar{x}_h a_{hj} x_j, \quad a_{jh} = \bar{a}_{hj}.$$

These resemble bilinear forms over $\mathcal{C}$ with coefficient matrix required to be Hermitian. There is, however, the added property that the second set of variables consists of the conjugates of the first. If the coefficient matrix happens to be real and the variables are restricted to being real, f reduces to a real quadratic form. Even without these specializations, however, the values of a Hermitian form are always real.

The proof of this property begins with the facts that each diagonal element a_{hh} is real and each product $\bar{x}_h x_h$ is real, so that all diagonal terms

$$a_{hh}\bar{x}_h x_h$$

are real. The remaining terms fall into pairs

$$a_{hj}\bar{x}_h x_j + a_{jh}\bar{x}_j x_h = a_{hj}\bar{x}_h x_j + (\overline{a_{hj}\bar{x}_h x_j}).$$

Each such pair is real because it is the sum of a complex number and its conjugate. This proves the assertion. (See Exercise 2 of this section for a shorter proof!)

THEOREM 5–13.   *The values of a Hermitian form are real.*

A nonsingular linear change of variables has the appearance $X = PY$, $X = \text{col}(x_1, \ldots, x_n)$, and necessitates the formula

(21) $$\bar{X} = \bar{P}\bar{Y},$$

where $\bar{X} = \text{col}\ (\bar{x}_1, \ldots, \bar{x}_n)$, $\bar{Y} = \text{col}\ (\bar{y}_1, \ldots, \bar{y}_n)$. To carry out this substitution we first write the form f above as a matrix product:

$$f = \bar{X}'AX = (\overline{PY})'APY = \bar{Y}'\bar{P}'APY.$$

The new coefficient matrix is Hermitian:

$$(\overline{\bar{P}'AP})' = (P'\bar{A}\bar{P})' = (\bar{P}'\bar{A}'P) = \bar{P}'AP.$$

This gives an analogue of Theorem 5–1.

THEOREM 5–14. *A nonsingular linear substitution replaces a Hermitian form with matrix A by a Hermitian form with matrix $\bar{P}'AP$, P nonsingular.*

If the complex matrices B and A satisfy an equation

$$B = \bar{P}'AP,$$

where P is nonsingular, we say that B is *conjunctive* to A, or *Hermitely congruent* to A. Hermitian congruence (or conjunction) is an RST relation.

EXERCISES

★1. Prove that Hermitian congruence is an RST relation.

2. Prove directly from $f = \bar{X}'AX$ that Hermitian forms are real-valued. Hint: Show that $\bar{f} = \bar{f}' = f$.

3. Let A and B be Hermitian matrices of the same size. Prove (1) that $A + B$ is Hermitian; (2) that AB is Hermitian if and only if $AB = BA$.

4. Show that if A is any rectangular complex matrix, $A\bar{A}'$ and $\bar{A}'A$ are Hermitian.

5. If A is Hermitian, show that A' and $\bar{A}$ are Hermitian, and also A^{-1} if A is nonsingular.

★6. If A is Hermitian, so is $c_m A^m + \cdots + c_1 A + c_0 I$ for arbitrary real scalars c_h and any positive integer m. Prove this fact.

7. Show that the range of values of a Hermitian form is not altered by nonsingular linear substitutions.

5–8 Reduction of Hermitian matrices.   The reduction theory for Hermitian matrices under conjunction is an almost perfect analogue of the theory in Sections 5–3 and 5–4 for symmetric matrices relative to congruence, and the proofs require only minor changes.

The analogue of Theorem 5–4 is readily discovered: B is Hermitely congruent to A if and only if $B = \bar{E}'_k[\cdots(\bar{E}'_1 AE_1)\cdots]E_k$, each pair E_h, $\bar{E}'_h$ representing an elementary column operation and the "conjunctively corresponding" row operation. That is, if E multiplies the jth column by c, $\bar{E}'$ multiplies the jth row by $\bar{c}$; if E adds c times column j to column k, $\bar{E}'$ adds $\bar{c}$ times row j to row k; if E interchanges column j with column k, $\bar{E}'$ interchanges row j with row k.

With this tool we may follow the proof of Theorem 5–5 in proving that every Hermitian matrix A is Hermitely congruent to a real diagonal matrix in which the number of nonzero diagonal elements is the rank of A. The major patchwork occurs in the construction of b_{hh} in (11), which is now twice the real part of a_{hj} and may be zero. If so, a_{hj} must be pure imaginary, $a_{hj} = ic \neq 0$. We then modify the construction of $B = (b_{hj})$ by adding i times column j to column h, and (necessarily) then $-i$ times row j to row h. This gives

$$b_{hh} = ia_{hj} - ia_{jh} = i(a_{hj} - a_{jh})$$
$$= i(ic - \overline{ic}) = i(ic + ic) = -2c \neq 0.$$

The remainder of the proof is valid without change, provided that "corresponding row transformation" is always interpreted in the conjunctive sense. The diagonal matrix B finally obtained is conjunctive to A, hence Hermitian. Since the diagonal of a Hermitian matrix is real, B must be real.

Suitable interchanges of rows and of columns replace B by a conjunctive (and congruent) matrix

$$D = \text{diag} (d_1, \ldots, d_r, 0, \ldots, 0),$$

in which

$$d_h > 0, \quad d_j < 0. \quad (h = 1, \ldots, p; j = p + 1, \ldots, r)$$

Exactly as in the proof of Theorem 5–6, D is Hermitely congruent to diag $(I_p, -I_{r-p}, 0)$.

THEOREM 5–15. *Every Hermitian matrix A of rank r is Hermitely congruent to a matrix*

$$B = \begin{bmatrix} I_p & 0 & 0 \\ 0 & -I_{r-p} & 0 \\ 0 & 0 & 0 \end{bmatrix}.$$

The integer p is uniquely determined by A.

The uniqueness proof for Theorem 5–6 is still valid. The integer p is called the *index* of the Hermitian matrix A. The test for Hermitian congruence then follows quickly:

THEOREM 5–16. *Two $n \times n$ Hermitian matrices are Hermitely congruent if and only if they have the same rank and index.*

Since the values of a Hermitian form $f = \bar{X}'AX$ are real, positive definiteness of f and of A may be defined exactly as for quadratic forms and symmetric matrices over $\mathfrak{R}$. (See Section 5-2.) The fact that nonsingular linear substitutions do not change the range of values of f implies the analogue of Theorem 5-8.

THEOREM 5-17. *An $n \times n$ Hermitian matrix of rank r and index p is positive semidefinite if and only if $p = r$, and positive definite if and only if $p = r = n$.*

Hereafter, when a matrix A is called positive semidefinite or definite, it will be understood without explicit statement that A is Hermitian (hence symmetric if it is real). The reader may then prove for himself the following extensions of Theorems 5-9 and 5-10.

THEOREM 5-18. *A complex matrix A is positive definite if and only if $A = \bar{P}'P$, where P is a nonsingular complex matrix. Moreover, if A is positive definite, $|A|$ and all principal subdeterminants are positive.*

EXERCISES

1. Prove Theorem 5-18.
2. Show that the index of a real symmetric matrix A, as defined by Theorem 5-6 and Definition 3, is the same as its index p as given by Theorem 5-15.

5-9 Skew-Hermitian matrices. A complex matrix $A = (a_{hj})$ is called *skew-Hermitian* in the case where

$$\bar{A}' = -A$$

or, what is equivalent, $\bar{a}_{jh} = -a_{hj}$ for all h and j. If A is skew-Hermitian and real, it is skew, and every real skew matrix is skew-Hermitian. If $B = \bar{P}'AP$ is Hermitely congruent to A, then B also is skew-Hermitian.

The Hermitian congruence of Hermitian matrices and the congruence of real Hermitian matrices are analogous theories, leading, as we have seen, to identical canonical sets. It is not true, however, that a like analogy holds for the Hermitian congruence of skew-Hermitian matrices and the congruence of real skew matrices. The latter leads to a canonical set having diagonal blocks

$$(22) \qquad \begin{bmatrix} 0 & 1 \\ -1 & 0 \end{bmatrix},$$

whence follows the property that a real skew matrix has even rank. But the matrices

$$(i), \quad \begin{bmatrix} i & 0 & 0 \\ 0 & 2i & 0 \\ 0 & 0 & 3i \end{bmatrix},$$

are skew-Hermitian and do not have even rank, and therefore cannot be Hermitely congruent to a direct sum of matrices (22).

This disappointment is compensated by the existence of a very simple theory of canonical sets for skew-Hermitian matrices A. It starts with the observation that the matrices $\pm iA$ are Hermitian. Let us consider $H = -iA$:

$$\bar{H}' = +i\bar{A}' = i(-A) = -iA = H.$$

Then there is a nonsingular complex matrix P such that

$$\bar{P}'HP = B = \begin{bmatrix} I_p & 0 & 0 \\ 0 & -I_{r-p} & 0 \\ 0 & 0 & 0 \end{bmatrix},$$

so that, using $iH = i(-i)A = A$, we find that

$$i\bar{P}'HP = \bar{P}'AP = iB,$$

$$(23) \qquad \bar{P}'AP = \begin{bmatrix} iI_p & 0 & 0 \\ 0 & -iI_{r-p} & 0 \\ 0 & 0 & 0 \end{bmatrix}.$$

Thus every skew-Hermitian matrix A is Hermitely congruent to a matrix (23) in which r is the rank of A and p is the index of $-iA$.

The fact that the diagonal elements in (23) are pure imaginaries should not be surprising, by virtue of Exercise 1 below. Also, a necessary and sufficient condition for Hermitian congruence of two skew-Hermitian matrices, to be expected once a canonical set has been found, is given in Exercise 2.

EXERCISES

1. Prove that if A is skew-Hermitian the diagonal elements of A are pure imaginaries or zero.

2. Show that two $n \times n$ skew-Hermitian matrices A and B are Hermitely congruent if and only if they have the same rank and $-iA$ and $-iB$ have the same index. Show also that the latter may be replaced by the requirement that iA and iB have the same index.

★3. Let A be a square matrix over $\mathcal{C}$. Show that A is uniquely expressible in the form $A = H + S$, where H is Hermitian and S is skew-Hermitian.

APPENDIX

5–10 Rank of a Hermitian matrix. The result presented in Theorem 4–11 shortens the work of determining the rank of a matrix by means of subdeterminants. For Hermitian or symmetric matrices the work may be shortened further.

THEOREM 5–19. *Let A be symmetric or Hermitian. Then A has rank r if and only if A has an $r \times r$ principal submatrix which is nonsingular, and no larger principal submatrix which is nonsingular.*

It thus suffices to examine only principal submatrices when seeking the rank of A. The number of submatrices to be examined may be reduced still further.

THEOREM 5–20. *Let A be symmetric or Hermitian. Then A has rank r if and only if A has a nonsingular $r \times r$ principal submatrix S, such that every principal submatrix of A containing S and one or two additional rows and columns is singular.*

THEOREM 5–21. *Let A be symmetric or Hermitian. If all principal submatrices having $r + 1$ rows or $r + 2$ rows are singular, the rank of A does not exceed r.*

For all three of these results see Reference 6, pp. 79, 80.

5–11 Positive definiteness. Theorem 9–26 (part of which is proved in Theorem 5–10) presents a criterion for positive definiteness of a Hermitian matrix A: $|A|$ and all principal subdeterminants shall be positive. The criterion is capable of considerable simplification. Let S_j denote the submatrix of A taken from the first j rows and columns.

THEOREM 5–22. *Let A be an $n \times n$ Hermitian matrix. Then A is positive definite if and only if all n submatrices S_j defined above have positive determinants.*

For proof see the article "On positive definite quadratic forms," by L. S. Goddard in *Publicationes Mathematicae*, Vol. 2 (1951), pp. 46, 47, or Reference 7, p. 138, or Reference 15, p. 91.

5–12 Congruent automorphs. If $P'SP = S$, the matrix P is called a *congruent* (or cogredient) *automorph* of S.

THEOREM 5–23. *Let S be a nonsingular, symmetric matrix over $\mathfrak{F}$. If K is any skew matrix over $\mathfrak{F}$ such that $(S + K)(S - K)$ is nonsingular,*

(24) $$P = (S + K)^{-1}(S - K)$$

is a congruent automorph of S.

The proof is given by the following computation, in which T denotes the matrix $S^{-1}K$:

$$
\begin{aligned}
P &= [S(I + T)]^{-1}S(I - T) = (I + T)^{-1}(I - T), \\
P' &= (S' - K')(S' + K')^{-1} = (S + K)(S - K)^{-1} \\
&= S(I + T)(I - T)^{-1}S^{-1}, \\
P'SP &= S(I + T)(I - T)^{-1}S^{-1}S(I + T)^{-1}(I - T) \\
&= S(I + T)[(I + T)(I - T)]^{-1}(I - T) \\
&= S(I + T)[(I - T)(I + T)]^{-1}(I - T) \\
&= S(I + T)(I + T)^{-1}(I - T)^{-1}(I - T) \\
&= S.
\end{aligned}
$$

THEOREM 5–24. *Let $\mathfrak{F}$ be a field in which $1 + 1 \neq 0$, and let S be a nonsingular, symmetric matrix over $\mathfrak{F}$. Then the matrices P of Theorem 5–23 constitute the totality of congruent automorphs of S such that $I + P$ is nonsingular.*

Each matrix P of the preceding theorem has the property that $I + P$ is nonsingular, for

$$
\begin{aligned}
I + P &= I + (S + K)^{-1}(S - K) \\
(S + K)(I + P) &= S + K + S - K = 2S.
\end{aligned}
$$

Since $2S = (S + K)(I + P)$ is nonsingular by hypothesis, $I + P$ also must be nonsingular.

Conversely, let $P'SP = S$ and let $I + P$ be nonsingular. We must find a skew matrix K such that $(S + K)(S - K)$ is nonsingular and (24) holds. The equation (24) is the clue to the matrix K. Consider the equation

(25) $$(S + K)P = S - K,$$

in which K is an unknown matrix. Then:

$$
\begin{aligned}
K + KP &= S - SP = S(I - P), \\
K(I + P) &= S(I - P), \\
\end{aligned}
$$
(26) $$K = S(I - P)(I + P)^{-1}.$$

The matrix (26) will be shown to fulfill our requirements. We regard (26) as definition of K, the preceding work merely showing how one might be led to this definition.   Since the steps are reversible, however, (26) leads to (25).   It remains only to prove (i) that K is skew and (ii) that $S + K$ is nonsingular, whence (25) will clearly imply the validity of (24).

To prove (ii) we compute as follows:

$$S + K = S[I + (I - P)(I + P)^{-1}]$$
$$= S[(I + P) + (I - P)](I + P)^{-1}$$
$$= S(2I)(I + P)^{-1} = 2S(I + P)^{-1}.$$

The nonsingularity of the factors on the right implies that of $S + K$. To prove (i) we compute as follows:

$$[(I + P)^{-1}]' = [(I + P)']^{-1} = (I + P')^{-1},$$
$$K' = (I + P')^{-1}(I - P')S$$
$$= (I + P')^{-1}(S - P'S)$$
$$= (I + P')^{-1}(S - SP^{-1}),$$

since

(27) $P'SP = S, \quad P'S = SP^{-1},$

the nonsingularity of P following from that of $S = P'SP$.   Then

$$K' = (I + P')^{-1}S(I - P^{-1})$$
$$= [S^{-1}(I + P')]^{-1}(I - P^{-1})$$
$$= (S^{-1} + S^{-1}P')^{-1}(I - P^{-1})$$
$$= (S^{-1} + P^{-1}S^{-1})^{-1}(I - P^{-1}),$$

since (27) implies that

$$S^{-1}P' = P^{-1}S^{-1}.$$

Thus

$$K' = [(I + P^{-1})S^{-1}]^{-1}(I - P^{-1})$$
$$= S(I + P^{-1})^{-1}(I - P^{-1})$$
$$= S(I + P^{-1})^{-1}P^{-1} \cdot P(I - P^{-1})$$
$$= S[P(I + P^{-1})]^{-1}(P - I)$$
$$= S(P + I)^{-1}(P - I)$$
$$= -S(I + P)^{-1}(I - P).$$

But $(I + P)^{-1}$ is a polynomial in $I + P$, hence in P, with scalar coefficients, so that $(I + P)^{-1}$ surely commutes with $I - P$.   The last equation above may thus be written as

$$K' = -S(I - P)(I + P)^{-1} = -K.$$

This completes the proof.

One application of congruent automorphs is to be found in Section 9–18. Another possible application may be discussed now in connection with the congruence of arbitrary square matrices. If we seek a canonical set of matrices to one of which every square matrix A is congruent, we might proceed as follows. Let

$$A = S + K,$$

where S is symmetric, K is skew, and both are $n \times n$. This decomposition is possible and unique (provided that $1 + 1 \neq 0$). Then K is congruent to a canonical matrix

$$Q'KQ = K_{ns} = \text{diag } (E_1, \ldots, E_s, 0), E_i = \begin{bmatrix} 0 & 1 \\ -1 & 0 \end{bmatrix},$$

as in Theorem 5–11, and

$$Q'AQ = A_1 = S_1 + K_{ns},$$

where $S_1 = Q'SQ$ is symmetric. Thus we may assume at the outset that $A = S + K_{ns}$. We now wish to perform congruences which do not disturb K_{ns}:

$$A = S + K_{ns} \to P'AP = P'SP + K_{ns}.$$

Thus we seek nonsingular congruent automorphs P of K_{ns}, and in the class of all such matrices P we wish to choose one such that $P'SP$ is as simple as possible. In short, the problem of finding canonical forms for arbitrary square matrices under congruence will be solved if it is solved for symmetric matrices relative to the restricted type of congruence

$$S \to P'SP,$$

in which P is a congruent automorph of K_{ns}.

In the case in which K is nonsingular a modified treatment of this problem is given by John Williamson in the paper, "On the algebraic problem concerning the normal forms of linear dynamical systems," *American Journal of Mathematics*, Vol. 58 (1936), pp. 141–163.

5–13 Maxima and minima. Let $f(x_1, \ldots, x_n)$ be a function of n real variables $x_1, \ldots, x_n$. For convenience we may let

$$x = (x_1, \ldots, x_n)$$

and write our function simply as $f(x)$. Let $c = (c_1, \ldots, c_n)$ be a vector of constants. Then $f(x)$ has a *relative minimum* at $x = c$ if $f(x) - f(c)$ is always positive when x is different from c but in some sufficiently small neighborhood of c. Similarly $f(x)$ has a *relative maximum* at $x = c$ if $f(x) - f(c)$ is always negative for x different from c but in some neighborhood of c.

It is interesting to note that quadratic forms may be associated with the problem of determining whether $f(x)$ has a relative minimum or maximum at $x = c$.

THEOREM 5-25. *Let* $f(x) = f(x_1, \ldots, x_n)$ *have partial derivatives up to and including those of order three, which are continuous in a neighborhood of* $x = c$. *Let the first partial derivatives of* $f(x)$ *be zero for* $x = c$, *and let*

$$f_{ij} = \frac{\partial^2 f}{\partial x_i \partial x_j}$$

evaluated at $x = c$. *Then* $f(x)$ *has a relative minimum (or relative maximum) at* $x = c$ *if the symmetric matrix* $A = (f_{ij})$ *is positive definite (or negative definite).*

Under the hypotheses it is known * that a necessary condition for a relative minimum or maximum at $x = c$ is that all first partial derivatives of $f(x)$ shall vanish at $x = c$. Then $f(x)$ has a Taylor's expansion about $x = c$ of the following type:

$$f(x) - f(c) = \Sigma h_i f_{ij} h_j + R,$$

where $h_i = x_i - c_i$ and R is a "remainder." When x is sufficiently close to c (that is, when all the differences $h_i = x_i - c_i$ are numerically $< d$ for some d which is suitably small) the remainder R is known to be negligible in comparison with

$$Q = \Sigma h_i f_{ij} h_j.$$

In particular the algebraic sign of $f(x) - f(c)$ is that of Q if all the h_i are sufficiently small. Thus $f(x)$ has a minimum at $x = c$ if the quadratic form Q assumes only positive values when $0 < |h_i| < d$, $i = 1, \ldots, n$. The latter property in turn is true if and only if Q is positive for *all* values of the h_i except $h_1 = \cdots = h_n = 0$. (Why?) Thus $f(x)$ has a minimum at $x = c$ if A is positive definite, and the "maximum result" follows similarly.

* See any text on advanced calculus.

CHAPTER 6

POLYNOMIALS OVER A FIELD

6–1 Polynomials. The notion that a polynomial is some expression with "many" terms is not only inaccurate, but also guilty of the fault that it entirely bypasses the main ideas associated with polynomials. Let $\mathfrak{F}$ be a field and x be an abstract symbol. Further symbols or "expressions" may be obtained by manipulating x and the quantities of $\mathfrak{F}$, subject to the usual rules of algebra. If in these manipulations with x and the quantities of $\mathfrak{F}$ we employ only the operations of addition, subtraction, and multiplication, all the symbols or expressions obtained are called *polynomials*, or, to be more complete, polynomials in x *over* $\mathfrak{F}$.

Thus $1 \cdot x = x$ is a polynomial in x. If $\mathfrak{F} = \mathfrak{R}$,

$$f(x) = \tfrac{1}{5}x + \sqrt{2}x^3$$

is a polynomial. Neither division nor the square root process have been used in constructing $f(x)$ since $\tfrac{1}{5}$ and $\sqrt{2}$ are in $\mathfrak{R}$; only multiplication and addition have been used. It is clear that any expression of the form

$$(1) \qquad f(x) = c_n x^n + c_{n-1}x^{n-1} + \cdots + c_1 x + c_0,$$

with all the c_i in $\mathfrak{F}$, is a polynomial. Conversely, every polynomial in x may be arranged in the form (1), so that polynomials over $\mathfrak{F}$ can be defined as all expressions of the form (1) with coefficients c_i in $\mathfrak{F}$. The polynomial (1) all of whose coefficients c_i are zero is called the *zero polynomial*.

The totality of polynomials in x with coefficients in $\mathfrak{F}$ is denoted by $\mathfrak{F}[x]$ (read "$\mathfrak{F}$ bracket x") and this set $\mathfrak{F}[x]$ is called a *polynomial domain* over $\mathfrak{F}$. If we abandon the notion of x as a variable, we may regard each polynomial $f(x)$ as a single, fixed entity, a member of the set $\mathfrak{F}[x]$. Regarded as a number system, $\mathfrak{F}[x]$ has many of the properties of a field as listed in Chapter 1 (all but M4). These amount to formal statements of well-known laws. The commutative and distributive laws say, respectively, that the equations

$$f(x)g(x) = g(x)f(x),$$
$$f(x)[g(x) + h(x)] = f(x)g(x) + f(x)h(x)$$

are universally true. In brief, the usual manipulations involving sums, differences, and products are valid in $\mathfrak{F}[x]$. The invalidity of postulate M4 means that not every $f(x)$ in $\mathfrak{F}[x]$ has an inverse in $\mathfrak{F}[x]$. Certainly for $f(x) = x$ we can find no $g(x)$ in $\mathfrak{F}[x]$ such that $xg(x) = 1$.

The quantities of $\mathfrak{F}$ are members of $\mathfrak{F}[x]$ and are called *constants* or constant polynomials. The nonzero constants are the only quantities of $\mathfrak{F}[x]$ which have inverses in $\mathfrak{F}[x]$. The zero constant and the zero polynomial are the same quantity of $\mathfrak{F}[x]$.

If the coefficient c_n of $f(x)$ in (1) is not zero, $f(x)$ is said to have *degree n*, and we write

$$n = \deg f(x)$$

to indicate this degree. This definition implies that the degree of a nonzero constant is zero. (Why?) The degree of the zero polynomial is not defined at all. It then follows that if $f(x)$ and $g(x)$ *have* degrees, say r and s, $f(x)g(x)$ has degree $r + s$:

$$(2) \qquad \deg [f(x)g(x)] = \deg f(x) + \deg g(x).$$

This formula (2) has some consequences that are worthy of note. Suppose that $g(x) \neq 0$ and $f(x)g(x) = 0.$* It follows that $f(x) = 0$. For otherwise both $f(x)$ and $g(x)$ have degrees, and (2) implies that $f(x)g(x)$ has a degree, which is a contradiction.

Another consequence is the familiar process of cancelling a common factor (provided it is not zero!). Suppose that

$$h(x)g(x) = j(x)g(x), \quad g(x) \neq 0.$$

Then $[h(x) - j(x)]g(x) = 0$, so the property just proved implies that the polynomial in brackets is zero, whence $h(x) = j(x)$.

If $f(x)$ in (1) has degree n, the coefficient c_n is called the *leading coefficient* of $f(x)$. If its leading coefficient is 1, $f(x)$ is called *monic*.

6-2 Divisibility. Given polynomials $f(x)$ and $g(x)$ over $\mathfrak{F}$, we say that $g(x)$ *divides* $f(x)$ if

$$(3) \qquad f(x) = g(x)h(x)$$

for some $h(x)$ in $\mathfrak{F}[x]$. Also $g(x)$ is called a *factor* of $f(x)$ and (3) is a

---

* That is, $f(x)g(x)$ is the zero polynomial but $g(x)$ is not. Unless the context makes the contrary clear, an equation $h(x) = 0$ will indicate that $h(x)$ is the zero polynomial, and will not indicate that we are seeking scalars c such that $h(c) = 0$.

factorization of $f(x)$ *over* $\mathfrak{F}$, this latter phrase indicating that the factors $g(x)$ and $h(x)$ have coefficients in $\mathfrak{F}$. Thus over $\mathfrak{F} = \mathfrak{R}$, $x + 1$ and $x - 1$ both divide $x^2 - 1$. So does $3x + 3$, since

$$x^2 - 1 = (3x + 3)(\tfrac{1}{3}x - \tfrac{1}{3}),$$

both factors on the right lying in $\mathfrak{R}[x]$.

In general, if $g(x)$ divides $f(x)$ and c is any constant $\neq 0$, $c \cdot g(x)$ also divides $f(x)$, since

$$f(x) = g(x)h(x) = c \cdot g(x)[c^{-1}h(x)]$$

with $c^{-1}h(x)$ in $\mathfrak{F}[x]$. In particular

$$f(x) = c \cdot c^{-1}f(x),$$

so that every nonzero constant divides every polynomial. A factorization into two factors, of which one is a constant, will be called a *trivial* factorization.

The possible factorizations (3) of a polynomial depend on the constant field $\mathfrak{F}$. Suppose, for example, that $\mathfrak{F}$ is the rational number field. Then $f(x) = x^2 - 2$ has no factorizations other than trivial ones. But the same polynomial $x^2 - 2$ belongs to $\mathfrak{R}[x]$, which has a larger constant field than the rationals. Here we find that

$$x^2 - 2 = (x - \sqrt{2})(x + \sqrt{2}),$$

so that $x^2 - 2$ factors over $\mathfrak{R}$ but not over the rationals. Again, $x^2 + 1$ does not factor over $\mathfrak{R}$ but does factor over $\mathfrak{C}$:

$$x^2 + 1 = (x + \sqrt{-1})(x - \sqrt{-1}).$$

DEFINITION 1. A nonconstant polynomial over $\mathfrak{F}$ is called *irreducible over* $\mathfrak{F}$ if all of its factorizations into two factors over $\mathfrak{F}$ are trivial.

Thus $x^2 - 2$ is irreducible over the rationals but not over $\mathfrak{R}$, and $x^2 + 1$ is irreducible over $\mathfrak{R}$ but not over $\mathfrak{C}$. Irreducibility of a polynomial with coefficients in $\mathfrak{F}$ is a property which may be lost when a larger constant field is contemplated.

The familiar process of long division of polynomials is valid over an arbitrary field. The following theorem, known as the "division algorithm," is a careful formulation of the result of "dividing as far as possible."

THEOREM 6–1. *If $f(x)$ and $g(x)$ are polynomials over $\mathfrak{F}$ and $g(x) \neq 0$, there are unique polynomials $q(x)$ and $r(x)$ over $\mathfrak{F}$ such that*

(4) $f(x) = q(x)g(x) + r(x),$

(5) $r(x) = 0$ or $\deg r(x) < \deg g(x).$

Let

$$f(x) = a_m x^m + \cdots + a_1 x + a_0,$$
$$g(x) = b_n x^n + \cdots + b_1 x + b_0,$$

where $b_n \neq 0$. If $f(x) = 0$ or $\deg f(x) < \deg g(x)$, conditions (4) and (5) are fulfilled by $q(x) = 0$, $r(x) = f(x)$. Now suppose that $f(x)$ has degree $m \geqq n$. Then

$$f(x) - a_m b_n^{-1} x^{m-n} g(x) = f_1(x)$$

has lower degree than $f(x)$, or is 0. The expression

$$a_m b_n^{-1} x^{m-n}$$

may be recognized in connection with the usual process of long division: It is the first term of the quotient. Also, $f_1(x)$ is the polynomial found after the first subtraction in the long division process.

If the degree m_1 of $f_1(x)$ is $\geqq n$, and if the leading coefficient of $f_1(x)$ is denoted by c, we again subtract a multiple of the divisor $g(x)$:

$$f(x) - a_m b_n^{-1} x^{m-n} g(x) - c b_n^{-1} x^{m_1-n} g(x) = f_2(x).$$

If $\deg f_2(x) \geqq n$, the process is repeated. The important feature of this procedure is that

$$\deg f(x) > \deg f_1(x) > \deg f_2(x) > \cdots.$$

Writing $t = m - n$, we eventually find an expression of the form

$$f(x) - k_1 x^t g(x) - k_2 x^{t-1} g(x) - \cdots - k_t g(x) = r(x),$$

where $r(x) = 0$ or $\deg r(x) < n$. This $r(x)$ and

$$q(x) = k_1 x^t + k_2 x^{t-1} + \cdots + k_t$$

fulfill conditions (4) and (5).

For the uniqueness, suppose also that

$$f(x) = p(x)g(x) + s(x),$$

where $s(x)$ is either the zero polynomial or a polynomial of degree less than n. Then

(6) $[p(x) - q(x)]g(x) = r(x) - s(x).$

The two sides of (6) are different notations for the same polynomial $h(x)$. From the right side of (6)

(7) $h(x) = 0$ or $\deg h(x) < n.$

From the left side of (6) we see that if $p(x) - q(x)$ is not zero, $h(x)$ has degree $\geq n$. The only tenable conclusion is that $p(x) - q(x)$ and $h(x)$ are 0, so that

$$p(x) = q(x), \quad s(x) = r(x).$$

This asserts the uniqueness.

The polynomial $r(x)$ is called the *remainder* of $f(x)$ under division by $g(x)$. An interesting case is that in which $g(x)$ is a monic polynomial of degree one, $g(x) = x - c$. Then $f(x) = q(x) \cdot (x - c) + r$, where r is a constant,

$$r = f(x) - q(x) \cdot (x - c).$$

Replacing x by c gives $r = f(c)$.

COROLLARY 6-1A. *When $f(x)$ is divided by $x - c$, the remainder is $f(c)$.*

In application of Corollary 6-1A suppose that c is a *root* of $f(x)$, that is, $f(c) = 0$. Then

$$f(x) = q(x)(x - c) + f(c) = q(x) \cdot (x - c).$$

This gives another result.

COROLLARY 6-1B. *If c is a root of $f(x)$, $x - c$ is a factor of $f(x)$, and conversely.*

The converse reads: If $f(x) = q(x) \cdot (x - c)$, c is a root of $f(x)$. But this is obvious.

The division algorithm (Theorem 6-1) settles questions as to whether a given polynomial $g(x) \neq 0$ divides (or is a factor of) another given polynomial $f(x)$. In equation (4) $g(x)$ surely divides $f(x)$ if $r(x) = 0$, but not otherwise. For, if $g(x)$ does divide $f(x)$, then $f(x) = h(x)g(x) = h(x)g(x) + 0$. The uniqueness in Theorem 6-1 then implies that $q(x) = h(x)$ and $r(x) = 0$. This gives:

COROLLARY 6-1C. *If $g(x)$ divides $f(x)$, the remainder $r(x)$ in (4) is zero, and conversely.*

EXERCISES

1. If $f(x)$ and $g(x)$ are nonzero polynomials in $\mathfrak{F}[x]$, find a necessary and sufficient condition that they shall divide each other.

2. Express each of the polynomials $f(x)$ below in the form (4), taking $\mathfrak{F} = \mathfrak{R}$:

    (a) $f(x) = x^4 + 4x^3 + 4x^2 - 7$, $g(x) = x^2 + 1$;

    (b) $f(x) = x^4 - 3x^2 + 5x + 6$, $g(x) = x - 2$.

3. If $f(x)$ divides $g(x)$ and $h(x)$, show that $f(x)$ divides $g(x) \pm h(x)$.

4. Let $\mathcal{K}$ be a field containing the field $\mathfrak{F}$, and let $f(x)$ and $g(x) \neq 0$ belong to $\mathfrak{F}[x]$. Show that if $f(x) = g(x)h(x)$ with $h(x)$ in $\mathcal{K}[x]$, then $h(x)$ lies in $\mathfrak{F}[x]$.

5. Show that the process of cancelling a nonzero factor $g(x)$ from an equation $h_1(x)g(x) = h_2(x)g(x)$ may be justified by the uniqueness in the division algorithm.

6-3 Greatest common divisors. If a polynomial $g(x)$ divides both $f_1(x)$ and $f_2(x)$, $g(x)$ is a *common* divisor of $f_1(x)$ and $f_2(x)$.

DEFINITION 2. A monic polynomial $d(x)$ is called the *greatest common divisor* of $f_1(x)$ and $f_2(x)$ if

(a) $d(x)$ is a common divisor of $f_1(x)$ and $f_2(x)$,

and

(b) every common divisor of $f_1(x)$ and $f_2(x)$ is a divisor of $d(x)$.

If a polynomial $d(x)$ is found with properties (a) and (b), then $k \cdot d(x)$ will also have these properties for every constant $k \neq 0$. In particular, k may be chosen so that $k \cdot d(x)$ is monic. Thus (a) and (b) are the crucial properties to seek in constructing a greatest common divisor, the property of monicity being easily achievable.

THEOREM 6-2. *If $f_1(x)$ and $f_2(x)$ are polynomials over $\mathfrak{F}$, not both zero, they have a greatest common divisor $d(x)$ in $\mathfrak{F}[x]$. Moreover, $d(x)$ is unique and is expressible as*

$$(8) \qquad d(x) = p_1(x)f_1(x) + p_2(x)f_2(x),$$

where $p_1(x)$ and $p_2(x)$ belong to $\mathfrak{F}[x]$.

Within $\mathfrak{F}[x]$ there is a subset $\mathcal{S}$ consisting of all polynomials of the form

$$(9) \qquad a_1(x)f_1(x) + a_2(x)f_2(x)$$

with $a_1(x)$ and $a_2(x)$ varying over $\mathfrak{F}[x]$. Then $\mathcal{S}$ contains

$$f_1(x) = 1 \cdot f_1(x) + 0 \cdot f_2(x)$$

and likewise $\mathcal{S}$ contains $f_2(x)$.

Observe that (i) $\mathcal{S}$ is closed under addition and subtraction. That is, a sum or difference of two polynomials of the form (9) is again of the form (9), hence in $\mathcal{S}$. Also (ii) a multiple $q(x)s(x)$ of a member $s(x)$ of $\mathcal{S}$ by an arbitrary polynomial $q(x)$ of $\mathfrak{F}[x]$ is again in $\mathcal{S}$. Observe, lastly, that $\mathcal{S}$ contains at least one nonzero polynomial, since it contains $f_1(x)$ and $f_2(x)$.

Among all nonzero polynomials in S, let $d(x)$ be one of minimal degree. We shall prove that $d(x)$ divides every $s(x)$ in S. By Theorem 6–1

$$s(x) = q(x)d(x) + r(x),$$

where $r(x) = 0$, or $r(x)$ has degree less than the degree of $d(x)$. By (ii) $q(x)d(x)$ belongs to S, and by (i) $r(x) = s(x) - q(x)d(x)$ belongs to S. Then $r(x) = 0$, for otherwise its degree would violate the minimal property of the degree of $d(x)$. Thus $s(x) = q(x)d(x)$, so that $d(x)$ divides every member of S. In particular, then, $d(x)$ divides both $f_1(x)$ and $f_2(x)$ so that $d(x)$ has property (a). By definition it has the form (8). If a polynomial $d_0(x)$ also divides $f_1(x)$ and $f_2(x)$,

$$f_1(x) = h_1(x)d_0(x), \quad f_2(x) = h_2(x)d_0(x),$$

then by (8)

$$d(x) = [p_1(x)h_1(x) + p_2(x)h_2(x)]d_0(x),$$

so that $d_0(x)$ divides $d(x)$: this is (b). As remarked above, the monic property is readily achieved. The uniqueness of $d(x)$ is left to the reader. We shall use the notation

$$d(x) = \gcd [f_1(x), f_2(x)]$$

for the greatest common divisor of $f_1(x)$ and $f_2(x)$.

EXERCISES

1. Find the greatest common divisor of each pair of polynomials below and express it in the form (8), always taking $\mathfrak{F} = \mathfrak{R}$:
    (a) $x^4 + x - 2, x^2 - 1$;
    (b) $x^3 + x^2 + x + 1, x^2 + 1$;
    (c) $x^3 + x + 1, x^2 + 1$.
2. Prove the uniqueness of $d(x)$ in Theorem 6–2.
3. Prove that $d(x) = \gcd [f_1(x), f_2(x)]$ if and only if $d(x)$ is a monic, common divisor of $f_1(x)$ and $f_2(x)$, which is expressible in the form (8).
4. Prove that $\gcd [f_1(x), f_2(x)]$ is a monic, common divisor of the $f_i(x)$ having greater degree than any other monic, common divisor of the $f_i(x)$.
5. Referring to (8), prove that $d(x) = q_1(x)f_1(x) + q_2(x)f_2(x)$, where $q_1(x) = p_1(x) + \dot{m}(x)f_2(x)/dx$, $q_2(x) = p_2(x) - m(x)f_1(x)/dx$, and $m(x)$ is any polynomial in $\mathfrak{F}[x]$.
6. Referring to (8) and to Exercise 5, prove that the only pairs of polynomials $q_i(x)$ obeying $d(x) = q_1(x)f_1(x) + q_2(x)f_2(x)$ are of the type given by Exercise 5.
7. If $\gcd [f(x), g(x)] \neq 1$, prove the existence of nonzero polynomials $a(x)$

and $b(x)$ such that $a(x)f(x) + b(x)g(x) = 0$, deg $a(x) <$ deg $g(x)$, deg $b(x)$ $<$ deg $f(x)$.

6–4 Relatively prime polynomials. Two polynomials are called *relatively prime* if their gcd is a constant (necessarily $= 1$). The polynomials $f_1(x) = x$ and $f_2(x) = x^2 - 1$ of $\mathfrak{R}[x]$ are relatively prime. One way of showing this is to observe that

$$1 = xf_1(x) + (-1)f_2(x).$$

(How do you finish the argument that gcd $[x, x^2 - 1] = 1$?)

LEMMA 6–1. *Let $p(x)$ be irreducible over $\mathfrak{F}$ and let $f(x)$ be any polynomial over $\mathfrak{F}$. Then $p(x)$ either divides $f(x)$ or is relatively prime to $f(x)$.*

Let $d(x) = $ gcd $[p(x), f(x)]$, so that

$$p(x) = d(x)q(x), \quad f(x) = d(x)g(x).$$

Since the first of these factorizations must be trivial, either $d(x)$ or $q(x)$ is a constant. If $q(x)$ is a constant, $d(x) = cp(x)$ whence $f(x) = c \cdot p(x)g(x)$, so that $p(x)$ divides $f(x)$. If $d(x)$ is a constant, $p(x)$ and $f(x)$ are relatively prime.

LEMMA 6–2. *Let $f(x)$ divide $g(x)h(x)$ and be relatively prime to $g(x)$. Then $f(x)$ divides $h(x)$.*

There are polynomials $a(x)$ and $b(x)$ such that

$$1 = a(x)f(x) + b(x)g(x).$$

On multiplication by $h(x)$ this becomes

$$h(x) = a(x)h(x)f(x) + b(x)g(x)h(x) = [a(x)h(x) + b(x)q(x)]f(x),$$

where by hypothesis $g(x)h(x) = q(x)f(x)$. The last equation above shows that $f(x)$ divides $h(x)$.

As a corollary of these lemmas the reader may prove

LEMMA 6–3. *If $p(x)$ is irreducible over $\mathfrak{F}$ and divides $g(x)h(x)$, it divides at least one of $g(x)$ and $h(x)$.*

Consider $h(x) = (x - 1)(x + 1)^2$. Then both $p(x) = x - 1$ and $q(x) = (x - 1)(x + 1)$ divide $h(x)$ but their product does not. On the other hand, there are cases where a product of two divisors of $h(x)$ is another such divisor. For example, $f(x) = x - 1$ and $g(x) = x + 1$ divide the $h(x)$ given above, and $f(x)g(x)$ also divides $h(x)$.

LEMMA 6–4. *Let $f(x)$ and $g(x)$ be relatively prime. Then if $f(x)$ and $g(x)$ divide $h(x)$, so does the product $f(x)g(x)$.*

By hypothesis

$$h(x) = g_1(x)g(x) = f_1(x)f(x).$$

Then $f(x)$ divides $g_1(x)g(x)$, and is relatively prime to $g(x)$. By Lemma 6–2 $f(x)$ divides $g_1(x)$,

$$g_1(x) = g_2(x)f(x), \quad h(x) = g_2(x)f(x)g(x).$$

The conclusion is displayed by the last equation.

EXERCISES

1. Answer the question preceding Lemma 6–1. Prove that if there are polynomials $a_i(x)$ and $f_i(x)$ such that $a_1(x)f_1(x) + a_2(x)f_2(x) = 1$, then $f_1(x)$ and $f_2(x)$ are relatively prime.

2. If $a_1(x)f_1(x) + a_2(x)f_2(x) = g(x)$, where the $a_i(x)$ and the $f_i(x)$ are polynomials, and $g(x)$ is a monic polynomial, is $g(x)$ necessarily the greatest common divisor of $f_1(x)$ and $f_2(x)$? If in addition $g(x)$ is assumed to be a divisor of $f_1(x)$ and $f_2(x)$ can we conclude that $g(x) = \gcd [f_1(x), f_2(x)]$?

3. Prove Lemma 6–3.

4. Suppose that two polynomials over $\mathfrak{F}$ are irreducible and distinct. Prove that one of the polynomials is the product of the other by a constant, or else the two polynomials are relatively prime.

5. Devise a proof for Lemma 6–4 which is similar to the proof of Lemma 6–2. Hint: Express the relative primeness of $f(x)$ and $g(x)$ by a formula like (8).

6–5 Unique factorization. The problem of factoring a polynomial as far as possible is familiar in elementary algebra, though usually it is not revealed there that the answer depends on the number system employed.

THEOREM 6–3. *Every nonzero polynomial $f(x)$ over $\mathfrak{F}$ is expressible in the form*

$$(10) \qquad f(x) = cp_1(x) \cdots p_r(x),$$

where c is a nonzero constant and the $p_i(x)$ are monic, irreducible polynomials of $\mathfrak{F}[x]$. The expression (10) is unique apart from the order in which the factors $p_i(x)$ appear.

The formula (10) is sometimes called the "prime factorization" of $f(x)$. We begin by factoring out the leading coefficient c of $f(x)$: $f(x) = cf_0(x)$, where $f_0(x)$ is monic. If $f_0(x)$ is irreducible over $\mathfrak{F}$,

(10) is achieved with $r = 1$ and $p_1(x) = f_0(x)$. Otherwise there is a factorization

$$f_0(x) = g(x)h(x),$$

where $g(x)$ and $h(x)$ have lower degrees than $f_0(x)$. Moreover, $g(x)$ and $h(x)$ may be taken to be monic. Since further factorization of $g(x)$ and $h(x)$ would lead to factors of still lower degree, this process must lead ultimately to a set of monic irreducible factors.

Uniqueness is the main problem. Suppose that

$$f(x) = cp_1(x) \cdots p_r(x) = cq_1(x) \cdots q_s(x),$$

where the $p_i(x)$ and the $q_j(x)$ are monic irreducible members of $\mathfrak{F}[x]$. Since $q_1(x)$ divides the product

$$p_1(x)[p_2(x) \cdots p_r(x)] = p_1(x)h_1(x),$$

it divides $p_1(x)$ or $h_1(x)$ by Lemma 6–3. If $q_1(x)$ divides $h_1(x)$ the argument may be repeated with $h_1(x) = p_2(x)[p_3(x) \cdots p_r(x)]$. Ultimately we find a $p_i(x)$ which $q_1(x)$ divides, and it is simply a matter of notation to take $i = 1$. Thus

$$p_1(x) = g(x)q_1(x).$$

Then $g(x)$ must be a constant c_1, since $p_1(x)$ is irreducible and $q_1(x)$ is not a constant. (Why is $q_1(x)$ not a constant?) Since $p_1(x)$ and $q_1(x)$ are monic, c_1 must be 1. Thus $p_1(x) = q_1(x)$, so that

$$p_2(x) \cdots p_r(x) = q_2(x) \cdots q_s(x)$$

by the cancellation process discussed at the end of Section 6–1. A repetition of the argument leads to $q_2(x) = p_2(x)$ and

$$p_3(x) \cdots p_r(x) = q_3(x) \cdots q_s(x).$$

Eventually the polynomials $q_j(x)$ or $p_i(x)$, say the former, are all cancelled off. If $s \neq r$, there would be some $p_i(x)$ left, and their product would equal 1. Since the $p_i(x)$ are nonconstants, this is impossible. Hence $s = r$ and every $q_i(x)$ is a $p_i(x)$. The uniqueness required in Theorem 6–3 is thus established.

Several of the $p_i(x)$ in (10) may coincide. If like factors are gathered to form a power of a single irreducible factor, Theorem 6–3 may be restated thus: Every polynomial $f(x)$ over $\mathfrak{F}$ is expressible uniquely, apart from order of the factors, as

(11) $$f(x) = cp_1(x)^{e_1} \cdots p_r(x)^{e_r},$$

where c is a constant and $p_1(x), \ldots, p_r(x)$ are *distinct*, monic polynomials which are irreducible over $\mathfrak{F}$.

The totality of monic divisors of $f(x)$ is easily described in terms of (11). It consists of all polynomials

$$(12) \qquad\qquad g(x) = p_1(x)^{a_1} \cdots p_r(x)^{a_r},$$

with each exponent a_i varying from 0 to e_i.

We frequently use the fact that if $f(x) = g(x)h(x)$, the prime factorization (10) of $f(x)$ can be found by substituting for $g(x)$ and $h(x)$ their prime factorizations.

EXERCISES

★1. Let $c(x)$ and $d(x)$ be relatively prime polynomials whose product is divisible by a monic polynomial $g(x)$. Prove that there are unique monic polynomials $c_0(x)$ and $d_0(x)$ such that $g(x) = c_0(x)d_0(x)$, $c_0(x)$ divides $c(x)$, and $d_0(x)$ divides $d(x)$.

2. List the totality of monic, common divisors of the following set of polynomials of $\Re[x]$:

$$(x^2 - 1)^2(x^2 + 1)(x + 2)^3, \quad (x^4 - 1)^2(x^2 - 4), \quad (x^2 + x - 2)^2.$$

6–6 Some applications. The statement that $m(x)$ is a *multiple* of $f(x)$ is by definition merely another way of saying that $f(x)$ divides $m(x)$. Then $m(x)$ is a common multiple of $f_1(x)$ and $f_2(x)$ in case it is a multiple of each of the $f_i(x)$. The *least common multiple* of $f_1(x)$ and $f_2(x)$ is a monic polynomial, which (a) is a common multiple of $f_1(x)$ and $f_2(x)$, and (b) has minimal degree. An equivalent definition not referring to degrees is obtained by replacing (b) by the requirement (b′) that $m(x)$ shall divide every common multiple of $f_1(x)$ and $f_2(x)$. We shall write

$$m(x) = \text{lcm } [f_1(x), f_2(x)]$$

to indicate that $m(x)$ is the least common multiple of the $f_i(x)$.

The prime factorization provided by Theorem 6–3 renders theoretically routine the calculation of the lcm and gcd of two polynomials. Suppose, for example, that $\Re$ is the coefficient field and that

$$f_1(x) = 5x^2(x - 1)^3(x^2 + 2), \quad f_2(x) = 6x^5(x - 1)(x + 1)^2.$$

Then x, $x - 1$, $x + 1$, and $x^2 + 2$ are all of the irreducible (over $\Re$) factors present. We may write

$$(13) \qquad \begin{aligned} f_1(x) &= 5x^2(x - 1)^3(x + 1)^0(x^2 + 2), \\ f_2(x) &= 6x^5(x - 1)(x + 1)^2(x^2 + 2)^0. \end{aligned}$$

Then since $d(x) = \text{gcd } [f_1(x), f_2(x)]$ is a divisor of each of the $f_i(x)$, it is by (12) a product of powers of x, $x - 1$, $x + 1$, and $x^2 + 2$:

$$d(x) = x^{a_1}(x - 1)^{a_2}(x + 1)^{a_3}(x^2 + 2)^{a_4}.$$

Each exponent a_i must be the smaller of the two corresponding exponents in (13), that is: a_1 must be the smaller of 2 and 5, a_2 the smaller of 3 and 1, a_3 the smaller of 0 and 2, and a_4 the smaller of 1 and 0. Thus

$$d(x) = x^2(x - 1).$$

If a larger exponent a_i were used for any i the polynomial $d(x)$ obtained would not divide both of the $f_i(x)$. If a smaller a_i were used, $d(x)$ would divide both $f_i(x)$, but would not be the "greatest" such divisor.

The lcm is constructed by the same process except that the larger exponent is used in each case. In the example,

$$\text{lcm } [f_1(x), f_2(x)] = x^5(x - 1)^3(x + 1)^2(x^2 + 2).$$

The generalization of both processes to arbitrary polynomials over arbitrary fields is immediate.

EXERCISES

1. Find the gcd and the lcm of each of the following pairs of polynomials in $\mathfrak{R}[x]$:
    (a) $x^3 - 1$, $x^2 - 1$;
    (b) $x^2 - 2$, $x^2 + 2\sqrt{2}x + 2$.
2. Prove that gcd $[f(x), g(x)] \cdot$ lcm $[f(x), g(x)] = f(x)g(x)$ for $f(x), g(x)$ monic.

6–7 Independence from $\mathfrak{F}$. Suppose that $f_1(x)$ and $f_2(x)$ are polynomials, not both zero, with coefficients in a field $\mathfrak{F}$. Then Theorem 6–2 provides a unique $d(x) = $ gcd $[f_1(x), f_2(x)]$ within $\mathfrak{F}[x]$.

However, if $\mathfrak{F}$ is contained in a larger field $\mathfrak{K}$, for example $\mathfrak{F} = \mathfrak{R}$ and $\mathfrak{K} = \mathfrak{C}$, then $f_1(x)$ and $f_2(x)$ also lie in $\mathfrak{K}[x]$ and may have many more divisors in $\mathfrak{K}[x]$ than in $\mathfrak{F}[x]$. It is conceivable that $f_1(x)$ and $f_2(x)$ may have within $\mathfrak{K}[x]$ a gcd, $D(x)$, which is different from their gcd, $d(x)$, found in the smaller system, $\mathfrak{F}[x]$. Actually this does not happen: $D(x) = d(x)$, as we shall see.

THEOREM 6–4. *If $f_1(x)$ and $f_2(x)$, not both zero, belong to $\mathfrak{F}[x]$, their gcd within $\mathfrak{F}[x]$ is also their gcd within $\mathfrak{K}[x]$, where $\mathfrak{K}$ is any field containing $\mathfrak{F}$.*

Let gcd $[f_1(x), f_2(x)]$ within $\mathfrak{F}[x]$ be

(14) $$d(x) = p_1(x)f_1(x) + p_2(x)f_2(x)$$

and let $D(x) = \gcd [f_1(x), f_2(x)]$ within $\mathcal{K}[x]$. Since $d(x)$ belongs to $\mathcal{K}[x]$ and is a common divisor of the $f_i(x)$, it certainly divides $D(x)$:

$$(15) \qquad\qquad D(x) = a(x)d(x).$$

But

$$f_i(x) = b_i(x)D(x), \qquad\qquad (i = 1, 2)$$

where the $b_i(x)$ are in $\mathcal{K}[x]$. Substituting in (14) we find

$$d(x) = [p_1(x)b_1(x) + p_2(x)b_2(x)]D(x) = b(x)D(x).$$

This result and (15) yield

$$D(x) = a(x)b(x)D(x).$$

Since $D(x)$ is not zero, $a(x)b(x) = 1$, so that the polynomials $a(x)$ and $b(x)$ must be nonzero constants. If $a(x) = $ constant $\neq 0$ is employed in (15), the fact that $D(x)$ and $d(x)$ are monic implies that $a(x) = 1$, $D(x) = d(x)$.

Thus the gcd of two polynomials depends only on the "smallest" field containing all of their coefficients. In particular the property that a pair of polynomials is relatively prime cannot be destroyed by enlargement of the coefficient field.

Exercises

1. List in the form (12) all the monic divisors of each of the polynomials $f(x) = (x^2 + 1)(x^2 + 2)$ and $g(x) = (x^2 + 1)^2(x^4 - 4)$, first within $\mathcal{R}[x]$, then within $\mathcal{C}[x]$. Finally, list their monic, common divisors in each case and thus determine gcd $[f(x), g(x)]$ in two ways.

2. Prove that the polynomials $x^2 + 2$ and $x + i$ of $\mathcal{C}[x]$ are relatively prime $(i = \sqrt{-1})$ by showing that $x^2 + 2$ and $x^2 + 1$ are relatively prime polynomials of $\mathcal{R}[x]$.

6–8 Generalizations. Until now we have discussed the gcd of only two polynomials. If $f_1(x), \ldots, f_r(x)$ belong to $\mathcal{F}[x]$, their gcd is defined as a monic polynomial $d(x)$ dividing all of the $f_i(x)$ and such that every common divisor of the $f_i(x)$ divides $d(x)$. That this gcd exists, provided at least one $f_i(x)$ is not the zero polynomial, may be established by a proof which is completely parallel to the proof given for Theorem 6–2. The same proof will show that

$$\gcd [f_1(x), \ldots, f_r(x)] = a_1(x)f_1(x) + \cdots + a_r(x)f_r(x)$$

for suitable polynomials $a_i(x)$ in $\mathcal{F}[x]$.

Theorem 6–4 and its proof may also be readily extended to more than two polynomials.

EXERCISES

★1. Let $f_1(x), \ldots, f_r(x)$ lie in $\mathfrak{F}[x]$. Using formulas like (12) for the totality of monic, common divisors in $\mathfrak{F}[x]$ of each of the $f_i(x)$, give a method for constructing the gcd of the $f_i(x)$.

★2. Prove the extension of Theorem 6–2 to the case of r polynomials $f_i(x)$.

★3. Prove the extension of Theorem 6–4 to the case of r polynomials $f_i(x)$.

4. Find the gcd and the lcm of the following set of four polynomials of $\mathfrak{R}[x]$:

$$(x^4 - 1)^2(x^2 + 4x + 4), \quad (x^2 - 1)(x^4 - 1),$$
$$(x^4 + 2x^2 + 1)(x^2 + 2), \quad (x^2 + 3x + 2)(x - 1)^2.$$

6–9 Roots of polynomials. If a nonconstant polynomial $f(x)$ over a field $\mathfrak{F}$ factors into linear factors,

$$(16) \qquad\qquad f(x) = c(x - r_1) \cdots (x - r_n),$$

we say that $r_1, \ldots, r_n$ are the *roots* of $f(x)$. The real polynomial $f(x) = (5x^2 + 10)^2$ does not have such a factorization over $\mathfrak{R}$, but does over the larger field $\mathfrak{C}$. There we have

$$(17) \qquad\qquad f(x) = 25(x + i\sqrt{2})^2(x - i\sqrt{2})^2,$$

and the roots of $f(x)$ are $i\sqrt{2}, i\sqrt{2}, -i\sqrt{2}$, and $-i\sqrt{2}$.

Suppose that $f(x)$ is a nonconstant polynomial over an arbitrary field $\mathfrak{F}$, such that $f(x)$ does not factor into linear factors over $\mathfrak{F}$. It is known that, as in the illustration above, there is a field $\mathfrak{K}$ containing $\mathfrak{F}$ such that $f(x)$ does have a factorization (16) with quantities r_i in $\mathfrak{K}$. See References 3, 4, and 17. The gist of the theory in these references is that roots of a polynomial are a well-established concept, and may be regarded as uniquely determined by the given polynomial. The reader will not go astray if he regards the general situation as similar to that in the example above with $\mathfrak{R}$ and $\mathfrak{C}$ as the fields.

If in (16) precisely k of the r_i are equal to r_1 we say that r_1 is a root of *multiplicity* k. In (17) there are four roots, but only two distinct roots, each having multiplicity two.

EXERCISES

1. Use the usual rules of calculus for the derivatives of products and polynomials to calculate the derivative $f'(x)$ of $f(x) = (x - r)^k g(x)$. Then prove that if r is a root of multiplicity k of $f(x)$, with $k > 1$, it is a root of multiplicity $k - 1$ of $f'(x)$.

2. Continue Exercise 1 to show that r is a root of $f(x)$ having multiplicity > 1 if and only if r is a common root of $f(x)$ and $f'(x)$.

CHAPTER 7

MATRICES WITH POLYNOMIAL ELEMENTS

7–1 Introduction. Many of the fundamental and practical results in the theory of matrices occur in the study of "similar matrices," treated in Chapter 8, and in related subjects. Though the matrices in question may have real number elements, the proofs of the theorems, and sometimes the theorems too, involve certain other matrices whose elements are polynomials. It is largely for these reasons that we have studied polynomials in the last chapter and will study matrices with polynomial elements in this chapter.

The main theme of this chapter appeared first in the theory of equivalence given in Chapter 3. A somewhat parallel development is given here for rectangular matrices with polynomial elements.

7–2 The status of past results. Let $\mathfrak{F}$ be a field and $\mathfrak{F}[x]$ the polynomial domain consisting of all polynomials in x with coefficients in $\mathfrak{F}$. We shall study rectangular matrices over $\mathfrak{F}[x]$, that is, with elements in $\mathfrak{F}[x]$.

Such concepts as nonsingularity, inverse, and rank were developed under the assumption that the scalars form a field. In the present situation the scalars are polynomials, and collectively constitute a polynomial domain $\mathfrak{F}[x]$, not a field. Fortunately, this does not lead to the abandonment of the concepts and results already developed, for, as we shall see, $\mathfrak{F}[x]$ is a part of a certain field $\mathfrak{K}$.

Consider all expressions obtained from x and the constants (elements of $\mathfrak{F}$) by use of addition, subtraction, multiplication, *and division*. This set of expressions is denoted by $\mathfrak{F}(x)$ and each of its members is called a *rational* expression in x with coefficients in $\mathfrak{F}$. As indicated in Chapter 1, $\mathfrak{F}(x)$ is a field, and by its definition $\mathfrak{F}(x)$ contains $\mathfrak{F}[x]$. Also, the familiar process of combining terms shows that every rational expression may be written as a single term

$$\frac{P(x)}{Q(x)},$$

whose numerator $P(x)$ and denominator $Q(x)$ belong to $\mathfrak{F}[x]$. In fact $\mathfrak{F}(x)$ may be defined as the totality of quotients of polynomials in x, denominators being nonzero polynomials.

Thus each matrix A over $\mathfrak{F}[x]$ is also a matrix with elements in the field $\mathfrak{K} = \mathfrak{F}(x)$, so that the past developments still apply.   To illustrate: The row space of A consists of all linear combinations of the rows of A with coefficients in the field $\mathfrak{K}$.   This space is a vector space over $\mathfrak{K}$ and its dimension is the rank of A.   The matrix A is nonsingular if and only if it is $n \times n$ and has rank n.   The determinant criterion for rank is, however, especially neat for matrices A over $\mathfrak{F}[x]$, since it involves only multiplication, addition, and subtraction, applied to the elements of A, hence may be computed entirely within $\mathfrak{F}[x]$.

EXERCISES

1. Find the inverses of the matrices

$$\begin{bmatrix} x & x+1 \\ x+2 & x+3 \end{bmatrix}, \quad \begin{bmatrix} x^2 & x+1 \\ x-1 & x^2 \end{bmatrix}.$$

2. By studying row spaces, show that the matrices

$$\begin{bmatrix} x & x+1 \\ x^2-x & x^2-1 \end{bmatrix}, \quad \begin{bmatrix} 0 & 1 & x \\ x & x & 1 \\ x^2-x & x^2-1 & x^2-1 \end{bmatrix}$$

are singular.   Show the same fact by proving that each matrix has zero determinant.

7-3 Equivalence over $\mathfrak{F}[x]$.   Elementary operations on a matrix over $\mathfrak{F}[x]$ are defined as follows:

   I. Interchange of two rows (or two columns).

   II. Multiplication of a row (or column) by a nonzero constant.

   III. Addition to the ith row of the product of the jth row by $f(x)$, where $j \neq i$ and $f(x)$ is any polynomial over $\mathfrak{F}$; and the analogous operation on columns.

These are precisely the operations of Chapter 2, but the word *scalar* now means a polynomial, and the scalars that have inverses are the nonzero constants.   As before, an elementary matrix E is a matrix obtained from I by performing a single elementary operation $\mathfrak{O}$ on I, and $\mathfrak{O}$ may be performed on A by multiplying by E on the appropriate side.

Let A and B be matrices over $\mathfrak{F}[x]$.   If B is obtained from A by performing any succession of elementary operations, B is said to be *equivalent* to A over $\mathfrak{F}[x]$.   As in the case of equivalence over $\mathfrak{F}$, B is equivalent to A over $\mathfrak{F}[x]$ if and only if $B = PAQ$, where P and Q are products of elementary matrices (in the sense defined above). When clearly understood, the phrase "over $\mathfrak{F}[x]$" is omitted.

It is evident from the definition that each elementary operation $\mathcal{O}$ on A may be "removed" by performing another elementary operation. When these two operations are performed on I we find

LEMMA 7–1. *Each elementary matrix over $\mathfrak{F}[x]$ has an inverse which is an elementary matrix over $\mathfrak{F}[x]$.*

As usual, each matrix is equivalent to itself; B equivalent to A implies A equivalent to B; and if B and A are equivalent to a common matrix, they are equivalent to one another. Thus, equivalence over $\mathfrak{F}[x]$ is another RST relation.

Since members of $\mathfrak{F}[x]$ also belong to the field $\mathcal{K} = \mathfrak{F}(x)$, if matrices B and A are equivalent over $\mathfrak{F}[x]$ they are also equivalent over $\mathcal{K}$ in the sense of equivalence defined in Chapter 3. The latter equivalence does not alter rank. This gives

LEMMA 7–2. *If A and B are equivalent over $\mathfrak{F}[x]$, they have the same rank.*

EXERCISES

1. Show that the matrices in Exercise 2, Section 7–2, are equivalent, respectively, to

$$\begin{bmatrix} 1 & 0 \\ 0 & 0 \end{bmatrix}, \quad \begin{bmatrix} 1 & 0 & 0 \\ 0 & 1 & 0 \\ 0 & 0 & 0 \end{bmatrix}.$$

Then verify Lemma 7–2 in these two cases.

2. For 3×3 matrices write the elementary matrix which adds to the third row the product of the first row by $x + 1$. Then find the inverse of this matrix.

3. Show that the matrices in Exercise 1, Section 7–2, are equivalent, respectively, to

$$\begin{bmatrix} 1 & 0 \\ 0 & 1 \end{bmatrix}, \quad \begin{bmatrix} 1 & 0 \\ 0 & x^4 - x^2 + 1 \end{bmatrix}.$$

7–4 The canonical set. The division algorithm (Theorem 6–1) is the main tool to be used in reducing all $n \times s$ matrices over $\mathfrak{F}[x]$ to a simple, canonical set under equivalence.

LEMMA 7–3. *Each nonzero matrix A over $\mathfrak{F}[x]$ is equivalent (over $\mathfrak{F}[x]$) to a matrix of the form*

$$B_1 = \begin{bmatrix} f(x) & 0 \\ 0 & A_1 \end{bmatrix},$$

where $f(x)$ is monic and has minimal degree among all nonzero elements of all matrices equivalent to A.

The class of all elements of all matrices equivalent over $\mathfrak{F}[x]$ to A contains a polynomial $f(x)$ of minimal degree. Since operations of type II are available, it may be assumed that $f(x)$ is monic. Let B denote a matrix equivalent to A and having $f(x)$ as one of its elements. Then $f(x)$ may be brought into the first row by an interchange of rows, and then into the first column by a column interchange. Hereafter we use only operations of type III beginning with the matrix C now at hand:

$$C = \begin{bmatrix} f(x) & C_1 \\ C_2 & C_0 \end{bmatrix}, \quad \begin{aligned} C_1 &= (c_{12}, \ldots, c_{1s}), \\ C_2 &= \operatorname{col}(c_{21}, \ldots, c_{n1}). \end{aligned}$$

Suppose that $c_{i1} = h(x) \neq 0$, $i > 1$. By the division algorithm

$$h(x) = q(x)f(x) + r(x),$$

where $r(x) = 0$, or $r(x)$ has degree less than that of $f(x)$. From the ith row of C we subtract the product of the first row by $q(x)$, thereby replacing c_{i1} by $r(x)$. The minimal property by which $f(x)$ was defined implies that $r(x) = 0$, so that c_{i1} has been replaced by 0. A repetition of this process makes the first column zero except for its top element $f(x)$. A similar treatment of the elements in the first row makes all of them zero except the $f(x)$, and this work does not alter the zeros already obtained in the first column. Hence the lemma.

THEOREM 7–1. *Each matrix of rank r with elements in $\mathfrak{F}[x]$ is equivalent over $\mathfrak{F}[x]$ to a matrix B such that* (a) *B is diagonal;* (b) *the first r diagonal elements of B are monic polynomials $f_1(x), \ldots, f_r(x)$;* (c) *the remaining diagonal elements, if any, are zero;* (d) *$f_i(x)$ divides $f_{i+1}(x)$, $i = 1, 2, \ldots, r - 1$.*

To prove this theorem, let the given matrix A be subjected to Lemma 7–3, which gives an equivalent matrix B_1. If the submatrix A_1 appearing in B_1 is nonzero it may be subjected to Lemma 7–3 by performing on B_1 elementary operations not involving the first row or column. This gives a matrix

$$B_2 = \begin{bmatrix} f_1(x) & 0 & 0 \\ 0 & f_2(x) & 0 \\ 0 & 0 & A_2 \end{bmatrix},$$

where $f_2(x)$ has minimal degree among all nonzero elements of all matrices equivalent to A_1. If A_2 is not zero, the process may be repeated. A finite number of repetitions of this process produces a matrix B with properties (a), (b), and (c) for some integer $r \geqq 0$. Since r is evidently the rank of B, it must (Lemma 7–2) be the rank of A. To prove (d) use the division algorithm to get

$$(1) \qquad f_{i+1}(x) = q_i(x)f_i(x) + r_i(x).$$

Then recall that at the ith stage in the construction of B we have a matrix

$$B_i = \text{diag } [f_1(x), \ldots, f_i(x), A_i],$$

where $f_i(x)$ has minimal degree among all elements of all matrices equivalent to A_{i-1}. One matrix equivalent to A_{i-1} is

$$C_i = \begin{bmatrix} f_i(x) & 0 & 0 \\ 0 & f_{i+1}(x) & 0 \\ 0 & 0 & A_{i+1} \end{bmatrix}.$$

To the second row of C_i we add $q_i(x)$ times the first row, then from the second column of this matrix subtract the first column, obtaining

$$\begin{bmatrix} f_i(x) & -f_i(x) & 0 \\ q_i(x)f_i(x) & r_i(x) & 0 \\ 0 & 0 & A_{i+1} \end{bmatrix}.$$

If $r_i(x)$ were not zero, it would have lower degree than $f_i(x)$ by the division algorithm. Since this is forbidden by the minimal property of $f_i(x)$, $r_i(x)$ must be zero. Then (1), with $r_i(x) = 0$, shows the truth of (d).

In the next section it will be shown that the matrix B of Theorem 7–1 is uniquely determined by A.

EXERCISES

For each of the following matrices find the corresponding matrix B of Theorem 7–1.

1. The first matrix in Exercise 2, Section 7–2.
2. The second matrix in Exercise 2, Section 7–2.
3. The matrix

$$\begin{bmatrix} x & x^2 \\ x^3 & x^5 \end{bmatrix}.$$

7–5 Invariant factors. Certain features of the subdeterminants of a matrix A over $\mathfrak{F}[x]$ are not altered by elementary operations. These will be investigated now.

LEMMA 7-4. *If P is a product of elementary matrices over $\mathfrak{F}[x]$, the $t \times t$ subdeterminants of PA are linear combinations, with coefficients in $\mathfrak{F}[x]$, of the $t \times t$ subdeterminants of A.*

It suffices to make the proof for the case in which P is an elementary matrix E, since linear combinations of linear combinations are again linear combinations. Let M be a $t \times t$ submatrix of PA and let S be the submatrix of A lying in the same position as M. Then S lies in certain rows of A, which will be said to *pass through S*. If E does not affect these rows, $M = S$. If E interchanges two of these rows, $|M| = - |S|$, and if it interchanges one of these rows with a row not passing through S, M is another submatrix of A, perhaps with one row out of order, so $|M|$ is plus or minus some subdeterminant of A. If E multiplies a row passing through S by c, $|M| = c|S|$. Suppose now that E adds to the ith row $f(x)$ times the jth. If row i does not pass through S, $|M| = |S|$, and this conclusion also holds when both row i and j pass through S by Theorem 4–7. There remains the case in which row i passes through S, but row j does not. Then

$$|M| = |S| + f(x)|T|,$$

where T is a submatrix of A or becomes one by moving one row to the appropriate position. This completes the proof.

LEMMA 7-5. *Let P and Q be products of elementary matrices. Then the gcd of all $t \times t$ subdeterminants of PAQ is also the gcd of all $t \times t$ subdeterminants of A.*

Let $d(x)$ be the gcd of all $t \times t$ subdeterminants of A, and $d_1(x)$ that of $B = PA$. Then $d(x)$ divides $d_1(x)$ by Lemma 7–4. Since also $A = P^{-1}B$, where P^{-1} is a product of elementary matrices, $d_1(x)$ divides $d(x)$. Both being monic, $d_1(x) = d(x)$. It remains to consider $B = PA$ and $C = BQ$. But $C' = Q'B'$, where Q' is a product of elementary matrices, so that the part already proved shows that the gcd of the $t \times t$ subdeterminants of C' is that of B'. The subdeterminants of C' are those of C (Theorem 4–3), and likewise for B' and B. This completes the proof.

LEMMA 7-6. *Let A be a nonzero matrix over $\mathfrak{F}[x]$ and B an equivalent matrix with the properties of Theorem 7–1. Then the gcd of all $t \times t$ subdeterminants of A is*

$$g_t = f_1(x) \cdots f_t(x),$$

for $t = 1, 2, \ldots, r$.

By Lemma 7–5 it suffices to show that g_t is the gcd of all $t \times t$ subdeterminants of B. All $t \times t$ subdeterminants of B are 0 except those with diagonals lying on the diagonal of B and with $t \leq r$. One such subdeterminant is g_t. The others are products

$$f_{k_1}(x) \cdots f_{k_t}(x) \tag{2}$$

with each $k_i \geq i$, so that $f_i(x)$ divides $f_{k_i}(x)$, and g_t divides the product (2). Thus g_t is the desired gcd.

We are now in position to prove the uniqueness of the matrix B in Theorem 7–1. Suppose that A is equivalent not only to B but also to C with the properties of Theorem 7–1. Let the r nonzero polynomials on the diagonal of C be $h_1(x), \ldots, h_r(x)$. Then by Lemma 7–6 the gcd of all $t \times t$ subdeterminants of A is

$$g_t = f_1(x) \cdots f_t(x) = h_1(x) \cdots h_t(x).$$

In particular $g_1 = f_1(x) = h_1(x)$ and

$$\frac{g_t}{g_{t-1}} = f_t(x) = h_t(x). \tag{3}$$

Thus

THEOREM 7–2. *The matrix B in Theorem 7–1 is uniquely determined by the given matrix A. The polynomials $f_t(x)$ on the diagonal of B are given by (3) where $g_0 = 1$ and g_k is the gcd of all $k \times k$ subdeterminants of A, $k = 1, 2, \ldots, r$.*

We thus have a canonical set of matrices called *Smith's canonical matrices* for the relation of equivalence over $\mathfrak{F}[x]$. The integer r in Theorem 7–1 is characterized as the rank of A, and now the polynomials $f_1(x), \ldots, f_r(x)$ are also characterized clearly as quotients of certain gcd's of subdeterminants of A.

The *invariant factors* of a matrix A over $\mathfrak{F}[x]$ are defined to be the polynomials $f_1(x), \ldots, f_r(x)$ appearing on the diagonal of the Smith canonical matrix equivalent to A.

THEOREM 7–3. *Two $n \times s$ matrices over $\mathfrak{F}[x]$ are equivalent over $\mathfrak{F}[x]$ if and only if they have the same invariant factors.*

If A_1 and A_2 have the same set of invariant factors, they are equivalent to the same Smith canonical matrix B, hence equivalent to each other. Conversely, if A_1 and A_2 are equivalent, and are equivalent to canonical matrices B_1 and B_2, respectively, the transitivity of equivalence implies that B_2 is equivalent to A_1. Thus two canonical matrices, B_1 and B_2, are equivalent to A_1. By the uniqueness $B_1 = B_2$, and this gives the conclusion.

COROLLARY 7-3. *Let $\mathcal{K}$ be a field containing $\mathcal{F}$, and let A_1 and A_2 be matrices over $\mathcal{F}[x]$. If A_1 and A_2 are equivalent over $\mathcal{K}[x]$, they are equivalent over $\mathcal{F}[x]$.*

Since the elements of A_1 and A_2 lie in $\mathcal{F}[x]$, the gcd's of the $t \times t$ subdeterminants of A_1 and A_2 also lie in $\mathcal{F}[x]$, and (Theorem 6–4) these gcd's do not change when A_1 and A_2 are regarded as matrices over the larger domain, $\mathcal{K}[x]$. The invariant factors, then, do not change, since they are ratios of these gcd's. Thus the equivalence of A_1 and A_2 over $\mathcal{K}[x]$ implies by Theorem 7–3 their equivalence over $\mathcal{F}[x]$.

EXERCISES

1. Find the gcd of all 2×2 subdeterminants of the second matrix in Exercise 2, Section 7–2, and of the equivalent matrix in Exercise 1, Section 7–3. Do the same for 1×1 subdeterminants.

2. Write the invariant factors of the matrix in the preceding exercise.

3. Without performing elementary operations, prove the equivalence of the matrices

$$\begin{bmatrix} x & x^2 \\ x^3 & x^5 \end{bmatrix}, \quad \begin{bmatrix} x & 0 \\ 0 & x^5 - x^4 \end{bmatrix}.$$

Then obtain the same result by elementary operations.

4. Find the invariant factors of the following matrix without calculating gcd's of subdeterminants:

$$\begin{bmatrix} x & x \\ x^2 + x & x \end{bmatrix}.$$

5. If two matrices over $\mathcal{F}[x]$ are equivalent over $\mathcal{K} = \mathcal{F}(x)$, are they necessarily equivalent over $\mathcal{F}[x]$?

6. Prove the converse part of Theorem 7–3 by the use of Lemma 7–5.

7. If A is a square matrix over $\mathcal{F}[x]$ prove that A is equivalent to its transpose.

7–6 Elementary matrices and inverses. If a square matrix M over $\mathcal{F}[x]$ is regarded as merely a special matrix with elements in the field $\mathcal{K} = \mathcal{F}(x)$, necessary and sufficient conditions for M to have an inverse are already known: $|M| \neq 0$ is one criterion. Others follow: M shall be $n \times n$ of rank n; the null space of M shall be the zero space; M shall have a right inverse or a left inverse.

A stronger question arises. Which matrices M over $\mathcal{F}[x]$ have inverses with elements in $\mathfrak{F}[x]$? We have already seen that this is a property of elementary matrices over $\mathfrak{F}[x]$, hence also of products of such matrices. We shall see (Theorem 7–5) that these are the only matrices with the property in question.

THEOREM 7–4. *A matrix M over $\mathfrak{F}[x]$ is a product of elementary matrices if and only if M is square and $|M|$ is a nonzero constant.*

If an elementary matrix E effects a row operation of types I, II, or III, its determinant is -1, a constant $c \neq 0$, or $+1$, respectively. Hence a product M of such matrices has a nonzero constant as its determinant. Conversely, let $|M|$ be a constant $k \neq 0$. By Theorem 7–1 there are products P and Q of elementary matrices such that $B = PMQ$ is a Smith canonical matrix,

$$B = \mathrm{diag}\,[f_1(x), \ldots, f_r(x), 0, \ldots, 0],$$

with invariant factors $f_1(x), \ldots, f_r(x)$. Then

(4) $$|B| = |P| \cdot |M| \cdot |Q| = k|P| \cdot |Q|.$$

By the opening remarks in this proof the quantities $|P| = p$ and $|Q| = q$ are nonzero constants, so that (4) gives $|B| = kpq \neq 0$. This proves that the rank r is equal to n, where B is $n \times n$, since otherwise $r < n$ and $|B| = 0$. Since

$$|B| = f_1(x) \cdots f_n(x) = kpq,$$

each invariant factor $f_i(x)$ is a constant, and being monic, $f_i(x) = 1$, $i = 1, \ldots, n$. Thus

$$B = I = PMQ, \quad M = P^{-1}Q^{-1},$$

where $P^{-1}Q^{-1}$ is a product of elementary matrices.

THEOREM 7–5. *A matrix M over $\mathfrak{F}[x]$ has an inverse with elements in $\mathfrak{F}[x]$ if and only if M is a product of elementary matrices, hence if and only if M is square and $|M|$ is a nonzero constant.*

When M is a product of elementary matrices, it has already been observed that M^{-1} exists and has elements in $\mathfrak{F}[x]$. Conversely, if M^{-1} exists and has elements in $\mathfrak{F}[x]$, its determinant also is a polynomial in x, and

$$MM^{-1} = I, \quad |M| \cdot |M^{-1}| = |I| = 1.$$

Thus $|M|$ and $|M^{-1}|$ are polynomials whose product is 1, and they must be nonzero constants. By Theorem 7–4, M is a product of elementary matrices. The final "if and only if" in Theorem 7–5 is merely a restatement of Theorem 7–4.

EXERCISES

1. Show that matrices B and A over $\mathfrak{F}[x]$ are equivalent if and only if there are matrices P and Q over $\mathfrak{F}[x]$ which have inverses over $\mathfrak{F}[x]$ such that $B = PAQ$.

2. Let M be a square matrix over $\mathfrak{F}[x]$. Show that M may be reduced to I by elementary row operations over $\mathfrak{F}[x]$ if and only if $|M|$ is a nonzero constant.

3. Prove that the following matrices over $\mathfrak{R}[x]$ are products of elementary matrices by finding their inverses:

$$A = \begin{bmatrix} x & x+1 \\ x+2 & x+3 \end{bmatrix}, \quad B = \begin{bmatrix} 1 & x \\ x & x^2+1 \end{bmatrix}.$$

Also express B as a product of two elementary matrices.

7–7 Matric polynomials. The elements a_{ij} of an $n \times s$ matrix A over $\mathfrak{F}[x]$ are polynomials

$$a_{ij} = c_t x^t + c_{t-1} x^{t-1} + \cdots + c_1 x + c_0, \qquad (c_t \neq 0)$$

where t and each c_k depend on i and j. If m denotes the maximum of the degrees t of the elements a_{ij} of A, the matrix A (assumed to be nonzero) can be written as a polynomial

$$(5) \qquad\qquad A = A_m x^m + \cdots + A_1 x + A_0,$$

whose coefficients A_k are $n \times s$ matrices over $\mathfrak{F}$ with $A_m \neq 0$. Then A is called a *matric polynomial* of degree m.

For example,

$$A = \begin{bmatrix} x^3 + 5x - 7 & x^4 + 2x^2 \\ x^2 + 3 & x^3 - 2x + 1 \end{bmatrix}$$

$$= \begin{bmatrix} 0 & 1 \\ 0 & 0 \end{bmatrix} x^4 + \begin{bmatrix} 1 & 0 \\ 0 & 1 \end{bmatrix} x^3 + \begin{bmatrix} 0 & 2 \\ 1 & 0 \end{bmatrix} x^2$$

$$+ \begin{bmatrix} 5 & 0 \\ 0 & -2 \end{bmatrix} x + \begin{bmatrix} -7 & 0 \\ 3 & 1 \end{bmatrix}.$$

This matric polynomial has degree 4. In general, each matrix A over $\mathfrak{F}[x]$ is expressible uniquely as a matric polynomial (5). The coefficient matrix A_k of x^k can be described as obtainable from $A = (a_{ij})$ by replacing each polynomial a_{ij} by the coefficient of x^k appearing in a_{ij}. The matrix A_m is called the *leading coefficient* of A, and if A_m is nonsingular A is called a *proper* matric polynomial. If A is not square it does not have the opportunity to be proper.

To indicate that A is a matric polynomial it is convenient to use the functional notation $A = A(x)$. Since the variable x in any matrix over $\mathfrak{F}[x]$ behaves like the constants in computations, it may be written in (5) on either side of the coefficient matrices. Thus,

$$\begin{bmatrix} 1 & 2 \\ 3 & 4 \end{bmatrix} x^2 = x^2 \begin{bmatrix} 1 & 2 \\ 3 & 4 \end{bmatrix} = x \begin{bmatrix} 1 & 2 \\ 3 & 4 \end{bmatrix} x.$$

Moreover, in each of the three products above, x may be replaced by any constant c without destroying the equalities. In a general matric polynomial $A = A(x)$, the matrix $A(c)$ obtained by writing c in place of x is well defined. The same matrix $A(c)$ is obtained if the replacement $x = c$ is made when A is written as in (5), or in the modified form having some or all of the factors x on the left, or if A is simply written as a single $n \times s$ matrix with polynomial elements.

When x is replaced by a matrix C the definition of $A(C)$ is not so clear. From this point on, let us restrict attention to the case in which A is square. Consider the simple matric polynomial

$$A = A(x) = A_1 x = \begin{bmatrix} 1 & 0 \\ 0 & 0 \end{bmatrix} x = x \begin{bmatrix} 1 & 0 \\ 0 & 0 \end{bmatrix}.$$

Substituting

$$C = \begin{bmatrix} 1 & 2 \\ 3 & 4 \end{bmatrix}$$

for x gives

$$A_1 C = \begin{bmatrix} 1 & 2 \\ 0 & 0 \end{bmatrix}, \quad CA_1 = \begin{bmatrix} 1 & 0 \\ 3 & 0 \end{bmatrix}.$$

Since $A_1 C \neq CA_1$ it is clear that the side on which x appears at the moment of the substitution $x = C$ is of importance, and that $A(C)$ is not well defined without some agreement on this matter.

For a general $n \times n$ matric polynomial (5), and an $n \times n$ constant matrix C, we define the right functional value $A_r(C)$ and the left functional value $A_l(C)$ as follows:

$$A_r(C) = A_m C^m + \cdots + A_1 C + A_0,$$
$$A_l(C) = C^m A_m + \cdots + CA_1 + A_0.$$

If $A(x)$ and $B(x)$ are two matric polynomials, $S(x) = A(x) + B(x)$ is perfectly well defined, and clearly

$$S_r(C) = A_r(C) + B_r(C), \quad S_l(C) = A_l(C) + B_l(C).$$

The product $P(x) = A(x)B(x)$ is also well defined, but the functional values $P_r(C)$ and $P_l(C)$ bear practically no relation to those of $A(x)$ and $B(x)$. To illustrate, let

$$A(x) = A_1x + A_0, \quad B(x) = B_1x + B_0,$$

so that

$$P(x) = A_1B_1x^2 + (A_0B_1 + A_1B_0)x + A_0B_0.$$

Then

$$P_r(C) = A_1B_1C^2 + (A_0B_1 + A_1B_0)C + A_0B_0,$$

whereas

$$A_r(C) = A_1C + A_0, \quad B_r(C) = B_1C + B_0,$$
$$A_r(C)B_r(C) = A_1CB_1C + A_1CB_0 + A_0B_1C + A_0B_0.$$

Since the matrix C, unlike the variable x, may not be carried to the right of B_1 and B_0 (except in very special cases), it is clear that the product of right functional values may not equal the right functional value of the product.

To repeat, there is no trouble when a scalar c (or a scalar matrix cI) is substituted for x; $P(c) = A(c)B(c)$ regardless of whether right or left functional values are used. The difficulties associated with substitution of a matrix for x, especially when multiplication of matric polynomials is involved, have been set forth both as a caution and as a background against which the results of the next section may appear in their proper significance.

EXERCISES

1. For the matrix

$$A = A(x) = \begin{bmatrix} 1 & x+1 \\ x^2 & 1+x+x^2 \end{bmatrix},$$

compute $A(c)$, $A_r(c)$, and $A_l(c)$, where c is an arbitrary scalar.

2. For the matric polynomial $A(x)$ in Exercise 1 compute $A_r(C)$ and $A_l(C)$, where

$$C = \begin{bmatrix} 1 & 2 \\ -1 & 0 \end{bmatrix}.$$

3. If $A(x)$ is the matric polynomial in Exercise 1 and

$$B(x) = \begin{bmatrix} x & 1 \\ 2 & -x \end{bmatrix}, \quad C = \begin{bmatrix} 1 & 0 \\ 0 & 2 \end{bmatrix},$$

compute $A_r(C)B_r(C)$ and $P_r(C)$, where $P(x) = A(x)B(x)$.

★4. Let $p(x) = a(x)b(x)$, where $a(x)$ and $b(x)$ are polynomials over $\mathfrak{F}$. Prove that $p(C) = a(C)b(C)$, where C is any square matrix over $\mathfrak{F}$.

★5. If A and B are $n \times n$ proper matric polynomials of degrees r and s, respectively, and D is any $n \times n$ matric polynomial $\neq 0$, show that the matric polynomial ADB has degree at least $r + s$.

7–8 Division algorithm for matric polynomials. The division algorithm of Theorem 6–1 concerns polynomials with coefficients in a field. An analogue will now be found for square matric polynomials.

THEOREM 7–6. *Let A and B be $n \times n$ matrices over $\mathfrak{F}[x]$ and let B be proper of degree s. Then there exist unique matrices Q, R, Q_1, and R_1 over $\mathfrak{F}[x]$, such that*

(6) $$A = QB + R, \quad A = BQ_1 + R_1,$$

where either $R = 0$ or the degree of R is less than s, and either $R_1 = 0$ or the degree of R_1 is less than s.

The proof is entirely analogous to that of Theorem 6–1, careful attention now being given to the noncommutativity of the coefficients. In the case of Theorem 6–1 the inverse of the leading coefficient of the divisor was used repeatedly. If

(7) $$B = B_s x^s + \cdots + B_1 x + B_0,$$

the hypothesis here that B is proper implies that B_s has an inverse. If $A = 0$ or if A has degree $m < s$, equations (6) are fulfilled by the matrices $R = R_1 = A$, $Q = Q_1 = 0$. Hence we let A be nonzero,

$$A = A_m x^m + \cdots + A_1 x + A_0 \qquad (A_m \neq 0)$$

and let $m \geqq s$. Then

(8) $$A - A_m B_s^{-1} B x^{m-s} = C_1$$

is zero or has degree $m_1 < m$. If C_1 is zero or of degree less than s, the first equation in (6) is satisfied by

$$R = C_1, \quad Q = A_m B_s^{-1} x^{m-s}.$$

If the degree of C_1 is $m_1 > s$, a finite repetition of the process gives the result exactly as in the case of Theorem 6–1. To get the second part of (6) alter (8) as follows:

$$A - B B_s^{-1} A_m x^{m-s} = D_1,$$

so that D_1 is zero or of lower degree than A. If D_1 is zero or of degree $m_1 < s$, we take

$$R_1 = D_1, \quad Q_1 = B_s^{-1} A_m x^{m-s}.$$

Otherwise we continue as above.   To prove the uniqueness, let

$$A = QB + R = Q_0B + R_0,$$

where R and R_0 individually are zero or of degree less than s.   Then

(9) $$(Q - Q_0)B = R_0 - R,$$

where $R_0 - R = 0$ or $R_0 - R$ has degree less than s.   If $Q - Q_0$ is not zero, it is a matric polynomial of some degree t with a leading coefficient $Q_t \neq 0$, and B is a polynomial (7) with nonsingular leading coefficient B_s.   The left side of (9), when written as a matric polynomial, has leading coefficient $Q_tB_s \neq 0$.   Then the left side of (9) has degree $t + s \geqq s$, in conflict with the situation on the right side. This proves that

$$Q - Q_0 = 0, \quad R_0 - R = 0,$$

and gives the uniqueness.   A like proof holds for Q_1 and R_1.

The first formula in (6) is described as *division of A on the right by B;* Q is the *right quotient* and R is the *right remainder*.   The parallel "left" concepts are associated with the second formula in (6).

THEOREM 7-7.   *If A is the $n \times n$ matric polynomial*

$$A = A(x) = A_mx^m + \cdots + A_1x + A_0$$

and B is the $n \times n$ monic, linear matric polynomial $B = Ix - C$, the right and left remainders on division of A by B are

(10) $$R = A_mC^m + \cdots + A_1C + A_0 = A_r(C),$$
(11) $$R_1 = C^mA_m + \cdots + CA_1 + A_0 = A_l(C).$$

The factorization

(12) $$x^iI - C^i = (x^{i-1}I + x^{i-2}C + \cdots + xC^{i-2} + C^{i-1})(xI - C)$$

can be verified immediately by multiplying out the product on the right.   If both sides of (12) are multiplied on the left by A_i and the resulting equations summed for $i = 1, 2, \ldots, m$, there results on the right a product $Q \cdot (xI - C)$, where Q is a matric polynomial.   On the left there results

$$\sum_{i=1}^{m} A_ix^i - \sum_{i=1}^{m} A_iC^i = \left(A_0 + \sum_{i=1}^{m} A_ix^i\right) - \left(A_0 + \sum_{i=1}^{m} A_iC^i\right)$$
$$= A - A_r(C).$$

Thus

$$A = Q \cdot (xI - C) + A_r(C),$$

so that $R = A_r(C)$ by the uniqueness of R in Theorem 7–6. A similar proof, reversing the two factors on the right of (12), gives (11).

7–9 Consequences of the division algorithm. With each $n \times n$ matrix C over $\mathfrak{F}$ there is associated the matric polynomial

$$xI - C,$$

which is called the *characteristic matrix* of C. Its determinant is a polynomial over $\mathfrak{F}$,

$$(13) \qquad |xI - C| = f(x) = x^n + c_{n-1}x^{n-1} + \cdots + c_0,$$

called the *characteristic polynomial* or *function* of C, and $f(x) = 0$ is called the *characteristic equation* of C.

THEOREM 7–8. *Let C be an $n \times n$ matrix over a field $\mathfrak{F}$ and let (13) be its characteristic function. Then*

$$f(C) = C^n + c_{n-1}C^{n-1} + \cdots + c_0I = 0.$$

This result, called the Cayley-Hamilton theorem, is frequently summarized by the statement: Every (square) matrix satisfies its characteristic equation. Note, as in Section 3–15, that when x^i is replaced by C^i in (13), the constant term c_0 is replaced by c_0I (which may be regarded as replacing c_0x^0 by $c_0C^0 = c_0I$).

To prove the theorem consider the adjoint matrix of $xI - C$:

$$\text{adj } (xI - C) = Q.$$

This is a matrix over $\mathfrak{F}[x]$. By Theorem 4–6

$$Q \cdot (xI - C) = |xI - C| \cdot I = f(x)I,$$
$$(14) \qquad Q \cdot (xI - C) = Ix^n + c_{n-1}Ix^{n-1} + \cdots + c_0I.$$

The right side of (14) is a matric polynomial A of degree n. If $B = xI - C$, (14) may be written as $A = Q \cdot B + 0$. The uniqueness of R in Theorem 7–6 then gives $R = 0$, but Theorem 7–7 gives

$$R = C^n + c_{n-1}C^{n-1} + \cdots + c_0I = 0.$$

An immediate corollary of Theorem 7–8 is this: The minimum polynomial (see Section 3–15) of an $n \times n$ matrix over a field $\mathfrak{F}$ has degree $\leq n$.

The next result, another application of Theorem 7–6, will be an important tool in Chapter 8.

THEOREM 7-9. *Let*

$$M_1 = A_1x + B_1, \quad M_2 = A_2x + B_2$$

be $n \times n$ *matric polynomials of degree one over* $\mathfrak{F}[x]$ *and let* M_2 *be proper. Then* M_2 *is equivalent to* M_1 *over* $\mathfrak{F}[x]$ *if and only if there are nonsingular matrices* S_0 *and* T_0 *over* $\mathfrak{F}$ *such that*

$$(15) \qquad\qquad M_2 = S_0 M_1 T_0.$$

If (15) holds with the stated properties of S_0 and T_0, these matrices are products of elementary matrices over $\mathfrak{F}$, and the latter are always elementary matrices over $\mathfrak{F}[x]$. Thus M_2 is equivalent to M_1. Conversely, let M_2 be equivalent to M_1 over $\mathfrak{F}[x]$, so that

$$(16) \qquad\qquad M_2 = SM_1T,$$
$$(17) \qquad\qquad SM_1 = M_2T^{-1}, \quad M_1T = S^{-1}M_2,$$

where S, T, S^{-1}, and T^{-1} are products of elementary matrices over $\mathfrak{F}[x]$. There are matric polynomials Q_1 and Q_2 such that

$$(18) \qquad\qquad S = M_2Q_2 + S_0, \quad T = Q_1M_2 + T_0$$

by Theorem 7-6 with $A = S$ and T in turn, and $B = M_2$. Since M_2 has degree one, S_0 and T_0 have elements in $\mathfrak{F}$. In the computation ahead we shall make repeated use of (17) and (18) in the effort to obtain (20), the right side of which has M_2 as a factor both on the left and the right:

$$
\begin{aligned}
M_2 &= SM_1T = (M_2Q_2 + S_0)M_1T \\
&= S_0M_1T + M_2Q_2S^{-1}M_2 \\
&= S_0M_1(Q_1M_2 + T_0) + M_2Q_2S^{-1}M_2 \\
&= S_0M_1T_0 + (S - M_2Q_2)M_1Q_1M_2 + M_2Q_2S^{-1}M_2 \\
&= S_0M_1T_0 + (SM_1)Q_1M_2 + M_2(-Q_2M_1Q_1 + Q_2S^{-1})M_2,
\end{aligned}
$$
$$(19) \quad M_2 = S_0M_1T_0 + M_2(T^{-1}Q_1 - Q_2M_1Q_1 + Q_2S^{-1})M_2,$$
$$(20) \qquad\qquad M_2 - S_0M_1T_0 = M_2DM_2.$$

The matrix D, equal to the expression in parentheses in (19), is a matric polynomial — recall that S^{-1} and T^{-1} are products of elementary matrices over $\mathfrak{F}[x]$! If $D \neq 0$, the right side of (20) is a matric polynomial of degree at least two, since M_2 is proper of degree one. The left has degree one, at most. Hence $D = 0$, and $M_2 = S_0M_1T_0$. To complete the proof we must show that S_0 and T_0 are nonsingular. But

$$M_2 = S_0M_1T_0 = S_0A_1T_0x + S_0B_1T_0 = A_2x + B_2.$$

By the unique expressibility of matric polynomials in the form (5),

(21) $$A_2 = S_0 A_1 T_0.$$

Since A_2 is nonsingular, S_0 and T_0 must be nonsingular.

EXERCISES

1. Prove that S_0 and T_0 in (21) are nonsingular in each of the following ways: (a) by determinants; (b) by using A_2^{-1} to show that S_0 has a right inverse and T_0 a left inverse; (c) by a theorem on the rank of a product.

2. Prove that the matrix M_1 in Theorem 7–9 is proper.

3. Compute the characteristic polynomials of the matrices

$$\begin{bmatrix} 1 & 2 \\ 3 & 4 \end{bmatrix}, \quad \begin{bmatrix} 1 & 0 & 1 \\ -1 & 0 & 0 \\ 2 & 0 & -1 \end{bmatrix},$$

and verify the Cayley-Hamilton theorem for these matrices.

★4. Show that the characteristic polynomial of any $n \times n$ matrix $C = (c_{ij})$ is monic, has $(-1)^n |C|$ as its constant term, and has $-(c_{11} + \cdots + c_{nn})$ as the coefficient of x^{n-1}.

★5. Let $f(x)$ be the minimum polynomial of a matrix A over $\mathfrak{F}$, and let $g(x)$ be a polynomial over $\mathfrak{F}$. Prove that $g(A)$ is nonsingular if and only if $g(x)$ is relatively prime to $f(x)$.

6. If $r_1, \ldots, r_n$ are the roots of the characteristic polynomial $f(x)$ of A and $h(x)$ is a polynomial, prove that $|h(A)| = h(r_1) \cdots h(r_n)$. Hint: Write $h(x) = c(s_1 - x) \cdots (s_k - x)$, so that $h(A) = c(s_1 I - A) \cdots (s_k I - A)$.

★7. Let A be a square matrix over $\mathfrak{F}$ whose characteristic polynomial has roots $r_1, \ldots, r_n$. If $g(x)$ is in $\mathfrak{F}[x]$, prove that the roots of the characteristic polynomial of $g(A)$ are $g(r_1), \ldots, g(r_n)$. Hint: $h(x) = y - g(x)$ is a polynomial over the field $\mathfrak{F}(y)$, and $h(A) = yI - g(A)$. Apply Exercise 6 to this polynomial $h(x)$ and matrix $h(A)$.

8. Show that M_1 and M_2 in Theorem 7–9 are equivalent if and only if $xI + A_1^{-1}B_1$ and $xI + A_2^{-1}B_2$ have the same invariant factors. (A_1 is nonsingular by the result in Exercise 2.)

APPENDIX

7–10 Systems of linear differential equations. Consider a system

$$\frac{d^2 y_1}{dx^2} - \frac{dy_1}{dx} + \frac{dy_2}{dx} = 0,$$

$$\frac{dy_1}{dx} - y_1 + \frac{dy_2}{dx} + y_2 = 0,$$

of ordinary linear differential equations with constant coefficients, y_1

and y_2 being unknown real functions of a real variable x. In terms of the operator

$$D = \frac{d}{dx},$$

the system is written as

(21)
$$\begin{aligned}(D^2 - D)y_1 + Dy_2 &= 0, \\ (D - 1)y_1 + (D + 1)y_2 &= 0,\end{aligned}$$

and in matrix notation:

(22)
$$\begin{bmatrix} D^2 - D & D \\ D - 1 & D + 1 \end{bmatrix}\begin{bmatrix} y_1 \\ y_2 \end{bmatrix} = \begin{bmatrix} 0 \\ 0 \end{bmatrix}.$$

It is easy to see that an *equivalent* system (one having the same solutions) is obtained if we (i) interchange two equations, (ii) multiply an equation by a nonzero constant, or (iii) add to one equation the result of operating on any other equation with any operator $f(D) = a_0 + a_1 D + \cdots + a_r D^r$, the a_i being any real numbers.

Steps such as these would commonly be employed in solving the system. We subtract from the first equation the result of operating with D on the second, then interchange the two equations, next multiply the second equation by -1. This gives

(23)
$$\begin{aligned}(D - 1)y_1 + (D + 1)y_2 &= 0, \\ D^2 y_2 &= 0.\end{aligned}$$

Then by well-known methods of elementary differential equations, we obtain

$$\begin{aligned}y_2 &= c_1 + c_2 x, \\ (D - 1)y_1 &= -(D + 1)y_2 \\ &= -c_1 - c_2 x - c_2 \\ &= -(c_1 + c_2 + c_2 x), \\ \begin{cases} y_1 &= c_1 + 2c_2 + c_2 x + c_3 e^x, \\ y_2 &= c_1 + c_2 x. \end{cases}\end{aligned}$$

The procedure used in reducing (21) to (23) should seem strangely familiar to the reader of Chapter 7. The operators (elements of the 2×2 matrix in (22)) are polynomials in D with constant coefficients. Such operators "multiply" like ordinary polynomials. That is, if we wish to operate on the second equation in (21) with $f(D) = a_0 + a_1 D + \cdots + a_r D^r$, the result can be seen to be

$$f(D) \cdot (D - 1)y_1 + f(D) \cdot (D + 1)y_2 = 0.$$

Formally we have merely "multiplied" the equation by $f(D)$. Multiplication and addition of these operators obey the associative, commutative, and distributive laws. In fact the operators $f(D)$ form a polynomial domain $\mathfrak{R}[D]$ subject to the usual rules for manipulating polynomials. Moreover, the processes (i), (ii), and (iii) on the matrix of coefficients in (22) are precisely the elementary row operations on matrices over $\mathfrak{R}[D]$.

In general, consider a system

(24)
$$a_{11}y_1 + \cdots + a_{1s}y_s = k_1,$$
$$a_{21}y_1 + \cdots + a_{2s}y_s = k_2,$$
$$\vdots$$
$$a_{n1}y_1 + \cdots + a_{ns}y_s = k_n,$$

in which $k_1, \ldots, k_n$ are real functions of x with suitable properties. The a_{ij} are polynomials in D with real constants as coefficients; that is, the a_{ij} belong to $\mathfrak{R}[D]$. Let

$$A = (a_{ij}),$$
$$Y = \text{col }(y_1, \ldots, y_s),$$
$$K = \text{col }(k_1, \ldots, k_n),$$

so that the system becomes

$$AY = K.$$

Elementary row operations, such as those illustrated above, may be effected by use of a matrix P, which is a product of elementary matrices over $\mathfrak{R}[D]$, giving a new system

$$PAY = PK.$$

Elementary column operations may also be brought into play. Suppose that Q is a product of elementary matrices effecting certain desired column operations. Then make a substitution

$$Y = QZ, \quad Z = \text{col }(z_1, \ldots, z_s),$$

thereby introducing new unknown functions z_i. The system then becomes

$$PAQZ = PK = \text{col }(h_1, \ldots, h_n).$$

The entire theory of Chapter 7 is now available. We choose the matrices P and Q so that the matrix PAQ of the new system is diagonal, invariant factors of A appearing on the diagonal:

$$PAQ = \text{diag }[f_1(D), \ldots, f_r(D), 0, \ldots, 0].$$

The system of differential equations now has variables separated:

$$f_1(D)z_1 = h_1,$$

.

.

.

$$f_r(D)z_r = h_r,$$
$$0z_{r+1} = h_{r+1},$$

.

.

.

$$0z_n = h_n.$$

If the rank r of A is equal to n, all the equations are of the form $f_i(D)z_i = h_i$. If $r < n$, "everything" depends on $h_{r+1}, \ldots, h_n$; the system is consistent if and only if the known functions $h_{r+1}, \ldots, h_n$ are zero.

Assuming that the system is consistent, we solve the first r equations, each involving only one unknown function, by known methods. If $f_i(D)$ has degree m_i in D, the solution for z_i will contain a number of arbitrary constants equal to m_i. The functions $z_{r+1}, \ldots, z_n$ are arbitrary. We then find Y from the equation $Y = QZ$, a process involving only linear combinations, with constant coefficients, of the functions z_i and their derivatives.

In most cases the $n \times s$ matrix A is $n \times n$ and nonsingular, so that $|A|$ is not the zero polynomial in D. We can then show that *the total number of arbitrary constants appearing in the solution functions $y_1, \ldots, y_n$ will be the degree m of $|A|$ as a polynomial in D.* First, the diagonal matrix PAQ above is now

$$PAQ = \text{diag} \, [f_1(D), \ldots, f_n(D)],$$

where

$$\deg f_i(D) = m_i > 0. \qquad (i = 1, \ldots, n)$$

Then the degree of $|PAQ|$ is

$$m_1 + \cdots + m_n,$$

and this is also the total number of arbitrary constants in $z_1, \ldots, z_n$. However, $|P|$ and $|Q|$ are nonzero constants so that

$$|PAQ| = |P| \cdot |Q| \cdot |A|$$

has the same degree in D as $|A|$:

$$m = m_1 + \cdots + m_n.$$

This proves that the degree of $|A|$ equals the total number of arbitrary constants in the solution functions $z_1, \ldots, z_n$. But we must consider whether any constants are lost when we compute

$$Y = QZ,$$

Q being a product of elementary matrices over $\mathfrak{R}[D]$.

Consider the ith row Q_i of Q:

$$Q_i = (q_{i1}, \ldots, q_{in}).$$

Then

$$y_i = q_{i1}z_1 + \cdots + q_{in}z_n.$$

Recall that if $k \neq j$, the arbitrary constants appearing in z_k are entirely distinct from those in z_j. Hence the only way a constant appearing in, say, z_j can be absent from y_i is for the constant to be absent from $q_{ij}z_j$, where q_{ij} is an operator belonging to $\mathfrak{R}[D]$. If this happens, the constant can be "saved" only by having it appear in some other function

$$y_h = q_{h1}z_1 + \cdots + q_{hn}z_n.$$

This, in fact, is what happens. For if the contrary were true, the constant in question would be absent from all of the functions

$$(25) \qquad\qquad q_{1j}z_j, \ldots, q_{nj}z_j.$$

But the relevant differential equations theory shows that the constant appears in z_j only as the coefficient of a function, $g(x) \neq 0$, which satisfies the equation

$$(26) \qquad\qquad f_j(D)g(x) = 0.$$

Moreover, this function depends on a number r in such a way that (26) implies that $f_j(r) = 0$. The constant in question can be absent from all of the functions (25) only if

$$q_{1j}g(x) = 0, \ldots, q_{nj}g(x) = 0,$$

whence

$$q_{1j}(r) = 0, \ldots, q_{nj}(r) = 0.$$

It follows that $D - r$ is a factor of all the polynomials $q_{1j}, \ldots, q_{nj}$ constituting column j of Q. Then $|Q|$ has $D - r$ as a factor, in conflict with the fact that $|Q|$ is a nonzero constant.

This completes the proof that no constants are lost in computing the "y solutions," $Y = QZ$, from the "z solutions." It also completes the proof that if $n = s$ and $|A| \neq 0$, the solutions of the system (24) involve a number of arbitrary constants equal to the degree of $|A|$ as a polynomial in D.

CHAPTER 8

SIMILARITY

8–1 Introduction. If B and A are $n \times n$ matrices over $\mathfrak{F}$, B is said to be *similar* to A over $\mathfrak{F}$ if there is a nonsingular matrix P with elements in $\mathfrak{F}$ such that

$$B = P^{-1}AP.$$

Then also A is similar to B:

$$A = PBP^{-1} = (P^{-1})^{-1}BP^{-1}.$$

Thus the relation of similarity is symmetric, and it suffices to say merely that B and A are similar. It is a simple matter to complete the verification that similarity is an RST relation.

Reasons for studying similarity appear in the next two chapters in many forms; in particular, significant ideas in geometry and dynamics depend upon the notion of similarity. The appendix to this chapter presents an instance of similarity arising in the study of a system of differential equations. All of these applications are closely related to the linear transformations studied in Chapter 10.

Exercises

★1. Prove that similarity is an RST relation.

2. Let $B = P^{-1}AP$. Show that a matrix Q satisfies $Q^{-1}AQ = B$ if and only if $Q = RP$, where R is a nonsingular matrix commuting with A.

8–2 Similarity invariants. The theory of matric polynomials may now be applied to similarity.

THEOREM 8–1. *Let A and B be $n \times n$ matrices over $\mathfrak{F}$. Then A is similar over $\mathfrak{F}$ to B if and only if $xI - A$ and $xI - B$ have the same invariant factors, hence, if and only if $xI - A$ and $xI - B$ are equivalent over $\mathfrak{F}[x]$.*

If $P^{-1}AP = B$, where P is a nonsingular matrix over $\mathfrak{F}$,

$$P^{-1}(xI - A)P = xI - B,$$

so $xI - A$ and $xI - B$ are equivalent, hence have the same invariant factors (Theorem 7–3). Under the converse hypothesis, Theorem 7–9 applies, so that

$$S(xI - A)T = xI - B,$$

where S and T have elements in $\mathfrak{F}$. Since the coefficient matrices of a matric polynomial are unique, $ST = I$ and $SAT = B$. Then $S = T^{-1}$ and B is similar to A.

> **COROLLARY 8–1.** *Let A and B be $n \times n$ matrices over $\mathfrak{F}$, and let $\mathfrak{K}$ be a field containing $\mathfrak{F}$. If A and B are similar over $\mathfrak{K}$, they are similar over $\mathfrak{F}$.*

In short, two matrices are similar over any field containing all of their elements or they are not similar at all. The hypothesis implies that $xI - A$ and $xI - B$ are equivalent over $\mathfrak{K}[x]$. By Corollary 7–3 they are equivalent over $\mathfrak{F}[x]$, whence (Theorem 8–1) A and B are similar over $\mathfrak{F}$.

As we have seen in Section 7–9, the determinant of the $n \times n$ matrix $xI - A$ is a monic polynomial of degree n. Since $xI - A$ thus has nonzero determinant it has rank n, hence has n invariant factors $f_1(x), \ldots, f_n(x)$. Let P_1 and P_2 be products of elementary matrices over $\mathfrak{F}[x]$ such that $P_1(xI - A)P_2$ is a Smith canonical matrix (Theorems 7–1, 7–2). Then

$$P_1(xI - A)P_2 = \text{diag } [f_1(x), \ldots, f_n(x)],$$
$$f_1(x) \cdots f_n(x) = |P_1(xI - A)P_2|$$
$$= |P_1| \cdot |P_2| \cdot |xI - A|$$
$$= p_1 p_2 f(x),$$

where $p_i = |P_i|$, $i = 1, 2$, and $f(x) = |xI - A|$. Since $f(x)$ and all of the $f_i(x)$ are monic, $p_1 p_2 = 1$ and

$$(1) \qquad\qquad f(x) = f_1(x) \cdots f_n(x).$$

In words, the characteristic polynomial of a matrix A is the product of the invariant factors of $xI - A$.

As Theorem 8–1 indicates, the discussion of similarity of matrices will involve frequent reference to the invariant factors of their characteristic matrices. To simplify these references somewhat we now introduce a new term.

> **DEFINITION 1.** The *similarity invariants* of an $n \times n$ matrix A are defined to be the invariant factors of $xI - A$.

The similarity invariants which are constants are equal to unity and are called *trivial*. Those which are not constants are called *nontrivial*.

EXERCISES

1. Prove (1) by the description of the similarity invariants as ratios of certain greatest common divisors.

2. Show that similar matrices have the same (a) characteristic polynomial, (b) rank, and (c) determinant.

3. Show that every square matrix is similar to its transpose.

★4. Let $g(x)$ be any polynomial over $\mathfrak{F}$. Show that $g(P^{-1}AP) = P^{-1}g(A)P$.

5. Use the result of Exercise 4 to show that similar matrices have the same minimum polynomial.

6. Use Theorem 8-1 to prove the similarity of the following matrices, a, b, and c being arbitrary nonzero scalars:

$$\begin{bmatrix} 0 & a \\ 0 & b \end{bmatrix}, \quad \begin{bmatrix} 0 & 0 \\ c & b \end{bmatrix}.$$

8-3 The minimum polynomial. In Chapter 2 it is pointed out that every $n \times n$ matrix A over $\mathfrak{F}$ satisfies a polynomial equation. The Cayley-Hamilton theorem in Chapter 7 asserts that A satisfies a particular polynomial equation of degree n. It follows that the minimum polynomial $m(x)$ has degree at most n. That $m(x)$ divides the characteristic polynomial of A is a special case of the following result.

THEOREM 8-2. *Let A be a square matrix over $\mathfrak{F}$ and let $g(x)$ be any polynomial over $\mathfrak{F}$. Then $g(A) = 0$ if and only if the minimum polynomial $m(x)$ of A divides $g(x)$.*

Let $g(A) = 0$. By the division algorithm for polynomials over $\mathfrak{F}$, $g(x) = q(x)m(x) + r(x)$. Since all the coefficients here are scalars, this formula remains valid if A is substituted for x:

$$0 = g(A) = q(A)m(A) + r(A) = r(A).$$

If $r(x)$ is not zero, it is of lower degree than $m(x)$, whence the property $r(A) = 0$, just proved, is in conflict with the definition of $m(x)$. Hence $r(x) = 0$ and $m(x)$ divides $g(x)$. Conversely,

$$g(x) = q(x)m(x),$$
$$g(A) = q(A)m(A) = q(A) \cdot 0 = 0.$$

This completes the proof.

That similarity invariant of A which has highest degree is $f_n(x)$.

We shall see that $f_n(A) = 0$, and later that $f_n(x) = m(x)$. Let $g_t(x)$ denote the gcd of the $t \times t$ subdeterminants of $xI - A$, so that $g_n(x) = f(x) = |xI - A|$ and (Theorem 7–2)

$$(2) \qquad\qquad f(x) = g_{n-1}(x)f_n(x).$$

Also, $g_{n-1}(x)$ is the gcd of all elements of

$$(3) \qquad\qquad Q(x) = \text{adj } (xI - A)$$

by the very definition of these elements. Then

$$(4) \qquad\qquad Q(x) = g_{n-1}(x)D(x),$$

where the gcd of all elements of $D(x)$ must be unity. Substitution of (4) and (2) in $Q(x) \cdot (xI - A) = f(x)I$ gives

$$g_{n-1}(x)D(x)(xI - A) = g_{n-1}(x)f_n(x)I,$$
$$(5) \qquad\qquad D(x)(xI - A) = f_n(x)I.$$

Equation (5) is recognizable as an application of the matrix division algorithm of Theorem 7–6 with $A = f_n(x)I$ and $B = xI - A$. Since (5) displays a remainder $R = 0$, Theorem 7–7 implies that

$$f_n(A) = 0.$$

Then the minimum polynomial $m(x)$ divides $f_n(x)$ by Theorem 8–2.

THEOREM 8–3. *The minimum polynomial of a square matrix A is that similarity invariant of A which has highest degree.*

Since we have seen above that

$$(6) \qquad\qquad f_n(x) = h(x)m(x),$$

it remains to prove that $h(x) = 1$.

The division algorithm may be applied to the matric polynomials $m(x)I$ playing the role of A in Theorem 7–6, and $xI - A$ playing the role of B. Then

$$(7) \qquad\qquad m(x)I = C(x) \cdot (xI - A),$$

where $C(x)$ is a matric polynomial and $R = m_r(A) = 0$. Substituting (7) in (6) and the result in (5), we find

$$D(x) \cdot (xI - A) = h(x)C(x)(xI - A),$$

so that

$$(8) \qquad\qquad D(x) = h(x)C(x)$$

by the uniqueness of the quotient Q in Theorem 7–6. The equation (8) asserts that the polynomial $h(x)$ divides all the elements of the matrix $D(x)$. As observed following (4), the gcd of these elements is

1, so that $h(x)$ must be a constant, h, and (6) becomes

$$f_n(x) = h \cdot m(x).$$

The monic property of $f_n(x)$ and $m(x)$ then implies that $h = 1$, and this completes the proof of Theorem 8–3.

A square matrix A is called *idempotent* if $A^2 = A$. Such matrices are the subject of attention not only in Exercise 8 below but also in numerous other places in the remainder of the book. Another interesting class of square matrices A consists of those for which some power A^t is the zero matrix. Such matrices A are called *nilpotent*.

EXERCISES

1. Use Theorem 8–3 to show that similar matrices have the same minimum polynomial.

2. Let A be a square matrix over $\mathfrak{F}$, and let $\mathfrak{K}$ be a field containing $\mathfrak{F}$. Prove that the minimum polynomial of A over $\mathfrak{F}$ is also the minimum polynomial of A over $\mathfrak{K}$.

3. Prove that A is nilpotent if and only if its minimum (or characteristic) polynomial has the form x^r.

4. Show that the minimum polynomial of A is linear if and only if A is a scalar matrix.

5. Using the result of Exercise 4, prove that if A is a 2×2 nonscalar matrix, the minimum and characteristic polynomials of A are equal.

6. Prove that all n similarity invariants of the $n \times n$ zero matrix are equal to x. Do this by computing only the characteristic and minimum polynomials and reasoning from these results.

7. Find the similarity invariants of I_n without computing gcd's of subdeterminants.

8. Prove that A is idempotent if and only if one of the following conditions is satisfied: (a) $A = 0$; (b) $A = I$; (c) A has minimum polynomial $x^2 - x$.

8–4 Equal characteristic and minimum polynomials. Since the invariant factors of any matrix over $\mathfrak{F}[x]$, in particular of $xI - A$, all divide the one of highest degree, we now know that the characteristic polynomial $f(x)$ of A is the product of the minimum polynomial $m(x)$ by certain monic divisors of $m(x)$. The latter divisors are all 1 if and only if $m(x) = f(x)$.

LEMMA 8–1. *Let A have characteristic and minimum polynomials both equal* to $f(x)$, and B have characteristic and minimum poly-*

---

* A matrix with minimum and characteristic polynomials equal is sometimes called *nonderogatory*.

nomials both equal to $g(x)$. *If* $g(x) = f(x)$, B *and* A *are similar* (*and conversely*).

Since their characteristic polynomials are equal, B and A have the same size, $n \times n$. Both have $n - 1$ similarity invariants equal to 1, and one invariant equal to $f(x)$. The similarity of B and A follows from Theorem 8–1. The converse is an exercise for the reader.

For every monic polynomial $g(x)$ we shall see that there is a matrix having $g(x)$ as its only nontrivial similarity invariant.

A monic polynomial of degree t over $\mathfrak{F}$ may always be written in the form

$$(9) \qquad g = g(x) = x^t - a_{t-1}x^{t-1} - \cdots - a_1x - a_0.$$

Associated with this polynomial is the $t \times t$ matrix

$$(10) \qquad \begin{bmatrix} 0 & 1 & 0 & \ldots & 0 & 0 \\ 0 & 0 & 1 & \ldots & 0 & 0 \\ \cdot & \cdot & \cdot & & \cdot & \cdot \\ \cdot & \cdot & \cdot & & \cdot & \cdot \\ \cdot & \cdot & \cdot & & \cdot & \cdot \\ 0 & 0 & 0 & \ldots & 0 & 1 \\ a_0 & a_1 & a_2 & & a_{t-2} & a_{t-1} \end{bmatrix} = C(g),$$

which is designated by $C(g)$ to show its relation to (9), and which is called the *companion matrix* of g. If $t = 1$ this requires elucidation. Then $g = x - a$ and $C(g)$ is the 1×1 matrix $C(g) = (a)$. Companion matrices will appear prominently in the canonical matrices for the similarity relation.

THEOREM 8–4. *The characteristic and minimum polynomials of the companion matrix of* $g(x)$ *are both equal to* $g(x)$.

The characteristic matrix of $C(g)$ in (10) is

$$(11) \qquad \begin{bmatrix} x & -1 & 0 & \ldots & 0 & 0 \\ 0 & x & -1 & \ldots & 0 & 0 \\ \cdot & \cdot & \cdot & & \cdot & \cdot \\ \cdot & \cdot & \cdot & & \cdot & \cdot \\ \cdot & \cdot & \cdot & & \cdot & \cdot \\ 0 & 0 & 0 & \ldots & x & -1 \\ -a_0 & -a_1 & -a_2 & \ldots & -a_{t-2} & x - a_{t-1} \end{bmatrix}.$$

The determinant of (11) is the characteristic polynomial $f(x)$ of $C(g)$, and may be computed after adding to the first column of (11) certain multiples of later columns. Then to this first column add x times column 2, x^2 times column 3, ..., x^{t-1} times column t. There results a matrix whose first column is zero except for the bottom element which is precisely $g(x)$. Expanding according to the elements of this modified first column, we find

$$f(x) = (-1)^{t-1}g(x) \cdot |{-}I_{t-1}| = g(x).$$

It remains to prove that $g(x)$ is also the minimum polynomial of $C(g)$. Since (11) has one $(t-1){\times}(t-1)$ subdeterminant,

$$|{-}I_{t-1}| = (-i)^{t-1} = \pm 1,$$

the gcd of all subdeterminants of this size is $g_{t-1} = 1$. But (Theorem 7-2)

$$f_t(x) = \frac{g_t}{g_{t-1}} = \frac{f(x)}{g_{t-1}} = f(x),$$

so that $f(x) = f_t(x)$. But Theorem 8-3 then finishes the proof by asserting that $f_t(x)$ is the minimum polynomial.

EXERCISES

1. Show that the characteristic and minimum polynomials of a matrix A are equal if and only if A has only one nontrivial similarity invariant.

2. If the characteristic polynomial of a matrix is a product of distinct linear factors, prove that it is equal to the minimum polynomial. Is the converse true? Why?

3. Prove that the characteristic and minimum polynomials of an $n \times n$ matrix A are equal if and only if the gcd of all $(n-1) \times (n-1)$ subdeterminants of $xI - A$ is 1.

4. If $A = C(g)$ is the companion matrix of $g = g(x) = x^3 - ax^2 - bx - c$, calculate A^2 and A^3, and show by direct computation that $g(x)$ is the minimum polynomial of A.

5. If $A = C(g)$ for any monic polynomial $g(x)$, prove that the minimum and characteristic polynomials of A' are both equal to $g(x)$.

6. If B is the $n \times n$ matrix

$$B = \begin{bmatrix} c_0 & c_1 & \cdots & c_{n-1} \\ c_{n-1} & c_0 & \cdots & c_{n-2} \\ \cdot & \cdot & & \cdot \\ \cdot & \cdot & & \cdot \\ \cdot & \cdot & \cdot & \cdot \\ c_1 & c_2 & \cdots & c_0 \end{bmatrix},$$

find a companion matrix A such that $B = c_{n-1}A^{n-1} + \cdots + c_1A + c_0I$. Hence prove that B is nonsingular if and only if the polynomials $x^n - 1$ and $c_{n-1}x^{n-1} + \cdots + c_1x + c_0$ are relatively prime. Use Exercise 5 of Section 7-9.

8-5 Direct sums. The theory of similarity is interesting not only because a canonical set is obtainable over a general field, but also because of the variety of well-known canonical sets. All of these sets employ companion matrices and direct sums (see Sections 1-6 and 4-3).

It should be recalled from Chapter 1 that if

$$B = \text{diag}(B_1, \ldots, B_r), \quad C = \text{diag}(C_1, \ldots, C_r),$$

with C_i of the same size as B_i, $i = 1, \ldots, r$, then

$$BC = \text{diag}(B_1C_1, \ldots, B_rC_r).$$

Moreover, if $P = \text{diag}(P_1, \ldots, P_r)$ is nonsingular, its inverse is

$$P^{-1} = \text{diag}(P_1^{-1}, \ldots, P_r^{-1}).$$

Then

$$P^{-1}BP = \text{diag}(P_1^{-1}B_1P_1, \ldots, P_r^{-1}B_rP_r).$$

This proves the first part of the following lemma, the second part having a like proof.

LEMMA 8-2. *If B_i is similar to C_i, $i = 1, \ldots, r$, diag $(B_1, \ldots, B_r)$ is similar to diag $(C_1, \ldots, C_r)$. If B_i and C_i are equivalent matrices over $\mathfrak{F}[x]$, $i = 1, \ldots, r$, diag $(B_1, \ldots, B_r)$ is equivalent over $\mathfrak{F}[x]$ to diag $(C_1, \ldots, C_r)$.*

It is convenient to know that each direct sum

(12) $$B = \text{diag}(B_1, \ldots, B_r)$$

is similar to any matrix

(13) $$C = \text{diag}(B_{i_1}, \ldots, B_{i_r})$$

obtained by "shuffling" the blocks B_i. The subscripts $i_1, \ldots, i_r$ in (13) denote an arbitrary rearrangement or permutation of the natural arrangement $1, 2, \ldots, r$.

LEMMA 8-3. *A direct sum (12) over $\mathfrak{F}$ is similar to any rearrangement (13). A direct sum (12) over $\mathfrak{F}[x]$ is equivalent over $\mathfrak{F}[x]$ to any rearrangement (13).*

It suffices to consider a rearrangement C obtained by interchanging only two blocks in (12), since a suitable succession of such steps yields an arbitrary permutation of the blocks. Suppose, then, that C results from interchanging blocks B_i and B_j in (12). Consider an identity matrix of the same size as B in (12) and partitioned in exactly the same manner as B:

(14) $I = \text{diag } (I_{n_1}, \ldots, I_{n_r})$.

Let P denote the matrix constructed by the interchange of the ith and jth columns of blocks in (14). Then multiplication of B on the right by P accomplishes the interchange of the ith and jth columns of blocks in B, and multiplying on the left by P' does the same for rows. Thus $P'BP$ is the desired matrix C. Since $P'P = I$, however, we find that

$$P^{-1} = P', \quad P^{-1}BP = C,$$

completing the proof of the first statement in the lemma. Since $|P|$ certainly is a nonzero element of $\mathfrak{F}$, the last equation may be interpreted as equivalence over $\mathfrak{F}[x]$ and this suffices to prove the second statement in the lemma.

If $B = \text{diag } (B_1, \ldots, B_r)$, then

$$B^k = \text{diag } (B_1^k, \ldots, B_r^k),$$

whence, if $m(x) = m_0 + m_1 x + \cdots + m_n x^n$ is any polynomial over $\mathfrak{F}$, we find that

$$m(B) = \text{diag } [m(B_1), \ldots, m(B_r)].$$

This property of direct sums is used frequently, and in particular in the proof of the next lemma.

LEMMA 8-4. *Let B and C have minimum polynomials $g(x)$ and $h(x)$, respectively. Then the minimum polynomial of $D = \text{diag } (B, C)$ is the least common multiple of $g(x)$ and $h(x)$, hence is $g(x)h(x)$ if $g(x)$ and $h(x)$ are relatively prime.*

For any polynomial $m(x)$ over the scalar field $\mathfrak{F}$ it is true that

$$m(D) = \begin{bmatrix} m(B) & 0 \\ 0 & m(C) \end{bmatrix}.$$

Then $m(D) = 0$ if and only if both of the following equations are valid:

$$m(B) = 0, \quad m(C) = 0.$$

By Theorem 8-2 these equations are equivalent to the requirement that both $g(x)$ and $h(x)$ divide $m(x)$. The minimum polynomial must, then, be the monic polynomial of least degree which satisfies this requirement, and this is precisely the definition of lcm $[g(x),h(x)]$. The final statement of the lemma is clear.

LEMMA 8–5. *Let* $p_1(x)$, $\ldots$, $p_s(x)$ *be distinct, monic, irreducible polynomials over* $\mathfrak{F}$, *and let* A_j *be a matrix having minimum and characteristic polynomials both equal to a power*

$$(15) \qquad\qquad p_j(x)^{e_j}. \qquad\qquad (j = 1, \ldots, s)$$

Then $C = \text{diag} (A_1, \ldots, A_s)$ *has*

$$f(x) = p_1(x)^{e_1} \ldots p_s(x)^{e_s}$$

both as its minimum and its characteristic polynomial. Hence C *is similar to any matrix having* $f(x)$ *as both characteristic and minimum polynomial.*

If (15) has degree n_j, $|xI - C|$ is the product of the determinants

$$|xI_{n_j} - A_j| = p_j(x)^{e_j},$$

so that C has the alleged characteristic polynomial. Its minimum polynomial is also $f(x)$ by repeated application of Lemma 8–4. The final statement is an immediate application of Lemma 8–1.

EXERCISES

1. Prove the second part of Lemma 8–2.
2. State and prove the extension of Lemma 8–4 to the case $D = \text{diag} (B_1, \ldots, B_r)$.
3. Specify matrices $A_1, \ldots, A_s$ fulfilling the hypotheses of Lemma 8–5.
4. If $B = \text{diag} (B_1, B_2, B_3)$ and $C = \text{diag} (B_3, B_1, B_2)$, where all the B_i are $n \times n$ blocks, construct a matrix P such that $P^{-1}BP = C$.
5. Show how to make the repeated applications of Lemma 8–4 required in the proof of Lemma 8–5.

8–6 Prescribed invariants. We are in position to get the first canonical set (Theorem 8–6) for similarity, the canonical matrices displaying their similarity invariants in the form of companion matrices. In the process we discover another important fact (Theorem 8–5), namely, that we can always find an $n \times n$ matrix having arbitrarily prescribed similarity invariants $f_1(x), \ldots, f_n(x)$, subject only to the necessary conditions:

(i) $n =$ the sum of the degrees of the $f_j(x)$,
(ii) $f_j(x)$ divides $f_{j+1}(x)$, $j = 1, \ldots, n - 1$.

Condition (i) reflects the fact that the characteristic polynomial of a matrix is the product of its similarity invariants. Note that either all of these invariants are of degree one (hence equal), or else some of them are trivial, that is, equal to 1.

THEOREM 8–5. *Let $g_1(x)$, . . . , $g_r(x)$ be nonconstant monic polynomials over $\mathfrak{F}$ such that $g_i(x)$ divides $g_{i+1}(x)$, $i = 1$, . . . , $r - 1$, and let $C(g_i)$ denote the companion matrix of $g_i(x)$. Then the matrix*

$$B = \operatorname{diag}\,[C(g_1),\, \ldots,\, C(g_r)]$$

has $g_1(x)$, . . . , $g_r(x)$ as its nontrivial similarity invariants.

Let deg $g_i = n_i$, so that $C(g_i)$ is $n_i \times n_i$. Then $xI - B$ is the direct sum of the blocks

$$xI_{n_i} - C(g_i),$$

whence the characteristic polynomial $f(x)$ of B is

$$f(x) = g_1(x) \cdots g_r(x).$$

If deg $f(x) = n$, B has n similarity invariants, of which precisely $n - r$ must be 1 if the theorem is true. The theorem will be proved if $xI - B$ is proved equivalent over $\mathfrak{F}[x]$ to

(16) $D(x) = \operatorname{diag}\,[1,\, \ldots,\, 1,\, g_1(x),\, \ldots,\, g_r(x)].$

Since $xI_{n_i} - C(g_i)$ has g_i as its only nontrivial invariant factor, it is equivalent over $\mathfrak{F}[x]$ to

$$B_i = \operatorname{diag}\,[1,\, \ldots,\, 1,\, g_i(x)].$$

Then (Lemma 8–2) $xI - B$ is equivalent to

$$D_1(x) = \operatorname{diag}\,(B_1,\, \ldots,\, B_r).$$

By the shuffling process of Lemma 8–3, $D_1(x)$ is equivalent to $D(x)$ in (16). This shows that $xI - B$ is equivalent to $D(x)$, which is a canonical matrix for equivalence; its diagonal elements are the invariant factors of $xI - B$. Thus $g_1(x)$, . . . , $g_r(x)$ are the nontrivial similarity invariants of B.

The first canonical set for similarity follows quickly from Theorem 8–5. Let A be an $n \times n$ matrix whose nontrivial similarity invariants are $g_1(x)$, . . . , $g_r(x)$. The matrix B of Theorem 8–5 is $n \times n$ (why?) and has the same nontrivial similarity invariants as A. Hence both have $n - r$ invariants equal to 1. The fundamental criterion in Theorem 8–1 implies that B and A are similar. This proves

THEOREM 8–6. *Every square matrix is similar to the direct sum of the companion matrices of its nontrivial similarity invariants.*

It should be recognized that if A has only one nontrivial similarity invariant $f(x)$, this theorem does not provide a direct sum similar to A. Rather, it implies that A is similar to $C(f)$. This slight discrepancy is akin to others such as a "product" which in special cases may have only one factor.

The blocks $C(g_i)$ in Theorem 8–6 may be written in any order by authorization of Lemma 8–3. If we arrange them in a prescribed order, say in the order of increasing size,* there results a completely unique matrix similar to a given matrix. With such an agreement Theorem 8–6 furnishes a canonical set for similarity.

EXERCISES

1. By calculating invariant factors of $xI - A$ find the canonical matrix of Theorem 8–6 similar to the matrix

$$A = \begin{bmatrix} 1 & 1 & -1 \\ 0 & 0 & 0 \\ 1 & 1 & -1 \end{bmatrix}.$$

Then obtain the same result again, but this time prove without using Theorem 8–3 that x^2 is the minimum polynomial of A, and x^3 is the characteristic polynomial. From these facts obtain all the similarity invariants.

2. If A is nilpotent prove that its canonical matrix of Theorem 8–6 is diag $(U_{n_1}, \ldots, U_{n_r})$, where U_k denotes the matrix $C(x^k)$, r is the number of nontrivial similarity invariants of A, and $n_1 \geqq n_2 \geqq \cdots \geqq n_r$.

3. Find the similarity invariants of I_5 and write its canonical matrix of Theorem 8–6.

4. Show without considerations of rank that the canonical matrix of Theorem 8–6 similar to the zero matrix is the zero matrix. Use the result of Exercise 6, Section 8–3.

8–7 Indecomposability. If a matrix A is similar over $\mathfrak{F}$ to a direct sum of two or more matrices, A is said to be *decomposable* over $\mathfrak{F}$; otherwise A is *indecomposable* over $\mathfrak{F}$. We shall find necessary and sufficient conditions for indecomposability.

THEOREM 8–7. *If A has characteristic polynomial $f(x) = c(x)d(x)$, where $c(x)$ and $d(x)$ are relatively prime, A is similar to a direct sum diag (C, D), where C and D have characteristic polynomials $c(x)$ and $d(x)$, respectively.*

Let the nontrivial similarity invariants of A be $g_1(x), \ldots, g_r(x)$, where each $g_i(x)$ divides $g_{i+1}(x)$. Then A is similar to the matrix

---

* Blocks of the same size cause no difficulty, since they are identical. Why?

$$B = \text{diag } [C(g_1), \cdots, C(g_r)],$$

and

$$f(x) = c(x)d(x) = g_1(x) \cdots g_r(x).$$

Each irreducible factor of $g_i(x)$ divides $c(x)$ or $d(x)$, but not both. Hence we find (Exercise 1, Section 6–5) unique monic polynomials $c_i(x)$ and $d_i(x)$ such that $c_i(x)$ divides $c(x)$, $d_i(x)$ divides $d(x)$, and

$$g_i(x) = c_i(x)d_i(x). \qquad (i = 1, \ldots, r)$$

It is clear that $c_i(x)$ is relatively prime to $d_i(x)$ and that

$$c(x) = c_1(x) \cdots c_r(x), \quad d(x) = d_1(x) \cdots d_r(x).$$

Let C_i and D_i be the companion matrices

$$C_i = C(c_i), \quad D_i = C(d_i),$$

where $c_i = c_i(x)$ and $d_i = d_i(x)$. Then

$$S_i = \text{diag } (C_i, D_i)$$

has characteristic polynomial $c_i(x)d_i(x) = g_i(x)$. Its minimum polynomial is also $g_i(x)$ by Lemma 8–4. Thus S_i is similar to $C(g_i)$ and (Lemma 8–2) may be substituted for $C(g_i)$ in

$$B = \text{diag } [C(g_1), \ldots, C(g_r)].$$

Thus A is similar to

$$\text{diag } (C_1, D_1, \ldots, C_r, D_r).$$

By the shuffling process, A is similar to

$$\text{diag } (C_1, \ldots, C_r, D_1, \ldots, D_r) = \text{diag } (C, D),$$

where the characteristic polynomials of

$$C = \text{diag } (C_1, \ldots, C_r), \quad D = \text{diag } (D_1, \ldots, D_r)$$

are $c(x)$ and $d(x)$, respectively. The proof is complete.

THEOREM 8–8. *Let A be a square matrix over $\mathfrak{F}$. Then A is indecomposable over $\mathfrak{F}$ if and only if its minimum and characteristic polynomials are equal and have the form $p(x)^e$, where $p(x)$ is irreducible over $\mathfrak{F}$.*

Let A be indecomposable. If A had two nontrivial similarity invariants, Theorem 8–6 would provide a "decomposition" for A; that is, A would be similar to a direct sum, contrary to assumption. Thus there is only one nontrivial invariant, and the minimum and characteristic polynomials of A are the same polynomial $f(x)$. The factorization of $f(x)$ into irreducible factors may be written as

$$f(x) = p_1(x)^{e_1} \cdots p_s(x)^{e_s},$$

where $p_1(x)$, ..., $p_s(x)$ are distinct, monic, irreducible polynomials over $\mathfrak{F}$. If $s > 1$,

$$f(x) = p_1(x)^{e_1}q(x), \quad \gcd [p_1(x)^{e_1}, q(x)] = 1.$$

By Theorem 8–7, A would be decomposable. Hence $s = 1$, and $f(x)$ has the alleged form.

Conversely, suppose that the minimum and characteristic polynomials of A are both equal to $p(x)^e$, where $p(x)$ is irreducible over $\mathfrak{F}$. If A were similar to a direct sum

$$S = \begin{bmatrix} C & 0 \\ 0 & D \end{bmatrix},$$

the characteristic polynomial $p(x)^e$ of S would be the product of those of C and D. Thus the characteristic polynomials of C and D are of the forms

$$p(x)^c, \quad p(x)^d, \quad c + d = e,$$

where both c and d are less than e. Let b denote the maximum of c and d, and let

$$m(x) = p(x)^b, \quad b < e.$$

Then $m(C) = 0$ and $m(D) = 0$, since the minimum polynomials of C and D divide $m(x)$. Since $m(S) = \text{diag } [m(C), m(D)]$, we see that $m(S) = 0$. The minimum polynomial of S and of A thus has smaller degree than has $p(x)^e$, contrary to the hypothesis that $p(x)^e$ is the minimum polynomial. The assumption that A is similar to a direct sum thus leads to a contradiction which establishes the theorem.

EXERCISES

1. For each of the following polynomials $f(x)$ with coefficients in the rational number field $\mathfrak{R}_0$, consider the matrix $A = C(f)$, whose elements lie in $\mathfrak{R}_0$ and whose minimum and characteristic polynomials are equal to $f(x)$. Determine in each case whether A is indecomposable over $\mathfrak{R}_0$, $\mathfrak{R}$, or $\mathfrak{C}$.

    (a) $x^2 - 3x + 2$.

    (b) $x^2 - 2x + 3$.

    (c) $(x^2 - 2x + 3)^3$.

    (d) $(x^2 - 5)^3$.

    (e) $(x - 2)^5$.

2. If the minimum polynomial of A is the product of two relatively prime polynomials $c(x)$ and $d(x)$, prove that A is similar to the direct sum of two

matrices whose minimum polynomials are $c(x)$ and $d(x)$, respectively. Hint: This is a corollary of Theorem 8-7.

8-8 Elementary divisors. The canonical matrices

$$(17) \qquad B = \text{diag } [C(g_1), \ldots, C(g_r)]$$

of Theorem 8-6 are direct sums of blocks $C(g_i)$ which may be decomposable since g_i may have several distinct, irreducible factors. The matrices (17), however, may be used in conjunction with other results to produce a canonical set which is a direct sum of indecomposable matrices.

There is a more compelling reason for attempting to improve on Theorem 8-6. It is desirable to choose canonical sets so that for any given matrix A, the canonical matrix similar to A is at least as "simple" as A. Yet it is possible to find a diagonal matrix A whose canonical matrix of Theorem 8-6 is nondiagonal! This form of embarrassment never occurs for the canonical set of the next theorem.

Let A, and the similar matrix B in (17), have characteristic polynomial

$$f(x) = p_1(x)^{e_1} \cdots p_s(x)^{e_s},$$

where $p_1(x), \ldots, p_s(x)$ are distinct, monic polynomials which are irreducible over the scalar field $\mathfrak{F}$, and each e_j is a positive integer. Then the similarity invariants $g_1(x), \ldots, g_r(x)$, obeying the divisibility conditions

$$g_i(x) \text{ divides } g_{i+1}(x), \qquad (i = 1, \ldots, r-1)$$

are of the form

$$(18) \qquad g_i(x) = p_1(x)^{e_{i1}} \cdots p_s(x)^{e_{is}}. \qquad (i = 1, \ldots, r)$$

Note that some of the e_{ij} may be 0, but if e_{ij} is positive, then $e_{i+1,j}$ is also positive, since

$$(19) \qquad e_{i+1,j} \geqq e_{ij}.$$

The condition (19) simply expresses the divisibility property of the invariants.

Those polynomials

$$p_j(x)^{e_{ij}}$$

which appear in the similarity invariants (18) with nonzero exponents e_{ij} are called* the *elementary divisors* of A over $\mathfrak{F}$. They are powers

---

* Sometimes called the elementary divisors of $xI - A$.

of irreducible polynomials over $\mathfrak{F}$, and their product is the characteristic polynomial of A. Notice that the list of elementary divisors may include duplications. A matrix over $\mathfrak{R}$ may be 12×12 and have nontrivial similarity invariants as follows:

$$g_1(x) = (x - 1)(x + 1),$$
$$g_2(x) = (x - 1)^2(x + 1)(x^2 + 2),$$
$$g_3(x) = (x - 1)^2(x + 1)(x^2 + 2).$$

Then nine similarity invariants are 1, and the list of elementary divisors is

$$x - 1, \quad (x - 1)^2, \quad (x - 1)^2,$$
$$x + 1, \quad x + 1, \quad x + 1,$$
$$x^2 + 2, \quad x^2 + 2.$$

THEOREM 8–9. *Every square matrix over $\mathfrak{F}$ is similar to the direct sum of the companion matrices of its elementary divisors over $\mathfrak{F}$.*

To prove this theorem, it suffices to start with the matrix B in (17) similar to A. A typical block, or direct summand, in B is $C(g_i)$, where $g_i = g_i(x)$ is shown in (18). The elementary divisors arising from $g_i(x)$ are then

$$(20) \qquad\qquad p_1(x)^{e_{i1}}, \ldots, p_s(x)^{e_{is}},$$

where it is understood that those polynomials in (20) which are constants (those with $e_{ij} = 0$) are to be deleted from the list. Then the companion matrices

$$A_{ij} = C[p_j(x)^{e_{ij}}]$$

have minimum and characteristic polynomials equal to these elementary divisors. By Lemma 8–5 the direct sum of these companion matrices for any fixed value of i is similar to $C(g_i)$ and (Lemma 8–2) may be substituted for $C(g_i)$ in (17). Theorem 8–9 is thus established.

The elementary divisors of a matrix are obtainable by factoring the similarity invariants into products of powers of distinct, irreducible factors. Conversely, given the list of elementary divisors, the similarity invariants are easily constructible. Suppose, for example, the elementary divisors of a matrix over $\mathfrak{R}$ are those in the list above Theorem 8–9. This list may be rewritten as follows:

$$
(21) \qquad
\begin{array}{ccc}
(x - 1)^2, & x + 1, & x^2 + 2, \\
(x - 1)^2, & x + 1, & x^2 + 2, \\
x - 1, & x + 1. &
\end{array}
$$

From the original list we have selected the highest powers of all the different irreducible polynomials; these constitute the first row above. From the remaining list of five polynomials, the highest powers were again selected and arranged as the second row above. The next repetition of the process depleted the list. If the polynomials in each row are multiplied, the three products obtained are the nontrivial similarity invariants.

That the procedure outlined is universally valid can easily be seen by use of the divisibility property of the similarity invariants. Since the elementary divisors thus determine the similarity invariants, and conversely, Theorem 8–1 has the following corollary: Two matrices over $\mathfrak{F}$ are similar if and only if they have the same elementary divisors.

Attention is called to the properties stated in Exercises 5 and 8 below. An arbitrary set of powers of irreducible polynomials can occur as the elementary divisors of a matrix; and a matrix A is similar to a diagonal matrix if and only if all the elementary divisors of A are linear.

There is one point deserving special emphasis in connection with elementary divisors of a matrix over $\mathfrak{F}$. Suppose that A is a real matrix whose elementary divisors over $\mathfrak{R}$ are those listed in (21). Over the field $\mathfrak{C}$ the polynomial $x^2 + 2$ is no longer a power of an irreducible polynomial. Thus the list of elementary divisors of A over $\mathfrak{C}$ is not (21) but rather:

$$(22) \quad \begin{matrix} (x-1)^2, & x+1, & x+\sqrt{-2}, & x-\sqrt{-2}, \\ (x-1)^2, & x+1, & x+\sqrt{-2}, & x-\sqrt{-2}, \\ x-1, & x+1. \end{matrix}$$

The similarity invariants of A do not change as the scalar field is enlarged, and the canonical matrices of Theorem 8–6 are not affected by enlargement of the field. Since the elementary divisors are subject to change, so are the canonical matrices of Theorem 8–9. For the matrix A, whose elementary divisors over $\mathfrak{R}$ are listed in (21), the canonical matrix over $\mathfrak{R}$ is

$$\text{diag } (A_1, A_2, A_2, B, B, B, C, C),$$

where

$$C = \begin{bmatrix} 0 & 1 \\ -2 & 0 \end{bmatrix}, \quad A_2 = \begin{bmatrix} 0 & 1 \\ -1 & 2 \end{bmatrix},$$
$$B = (-1), \qquad A_1 = (1).$$

Over $\mathbb{C}$ the elementary divisors of the same matrix are listed in (22) and the canonical matrix is

$$\text{diag } (A_1, A_2, A_2, B, B, B, C_1, C_1, C_2, C_2),$$

where

$$C_1 = (-\sqrt{-2}), \quad C_2 = (\sqrt{-2}).$$

EXERCISES

1. Consider a matrix A over the rational number field $\mathfrak{R}_0$ having in each of the problems (a)–(d) the given list of nontrivial similarity invariants. In each case find the three canonical matrices of Theorem 8–9 similar to A over $\mathfrak{R}_0$, $\mathfrak{R}$, and $\mathbb{C}$, respectively.

    (a) $(x^2 - 2)^2, x^2 - 2.$
    (b) $(x^2 - 4)^2, x^2 - 4.$
    (c) $(x^2 + 4)^2, x^2 + 4.$
    (d) $(x^2 - 4)^2, (x - 2), x - 2.$

2. If a matrix A has the following list of elementary divisors, write the canonical matrix of Theorem 8–6 similar to A: $x - 2, (x - 2)^2, x^2 + 1.$

3. If A is an idempotent matrix different from I and 0, show that the minimum polynomial of A is $x^2 - x$, and that the characteristic polynomial has the form $(x - 1)^r x^s$. Then use Theorem 8–9 to prove that A is similar to

$$B = \begin{bmatrix} I_r & 0 \\ 0 & 0 \end{bmatrix}.$$

Thus the rank of A is the number of elementary divisors equal to $x - 1$.

4. Using the results of Exercise 3, give a construction (in terms of non-singular matrices) for all idempotent matrices over a given field $\mathfrak{F}$.

★5. Let $f_1(x), \ldots, f_s(x)$ be powers of monic polynomials which are irreducible over $\mathfrak{F}$. Prove that there exists a matrix whose elementary divisors are the $f_i(x)$. Hint: The similarity invariants are determined.

6. Let $A = \text{diag } (A_1, \ldots, A_s)$, where each A_i has characteristic and minimum polynomial equal to $f_i(x)$, a power of a polynomial $p_i(x)$ which is irreducible over the scalar field $\mathfrak{F}$. Use Exercise 5 to prove that the $f_i(x)$ are the elementary divisors of A over $\mathfrak{F}$.

★7. Let $A = \text{diag } (A_1, \ldots, A_s)$. Prove that all the elementary divisors of all the matrices A_i constitute the elementary divisors of A. Hint: Apply Theorem 8–9 to each A_i. Use Exercise 6.

★8. Use Theorem 8–9 and Exercise 7 to show that a matrix A is similar to a diagonal matrix if and only if all of the elementary divisors of A are linear.

9. (a) Let A be diagonal. Prove that the canonical matrix of Theorem 8–9

similar to A is A itself. (b) Find a diagonal matrix A such that its canonical matrix of Theorem 8–6 is nondiagonal.

8–9 The rational canonical set. Perhaps the most important of the canonical sets for similarity is the one to be obtained in this section. It is called the *rational canonical* set, and the companion matrices appearing in it are all of the form $C(p)$, where $p = p(x)$ is irreducible over the scalar field $\mathfrak{F}$.

We shall have use for a square matrix N, all of whose elements are zero except the one in the lower left-hand corner, which is 1. The symbol N will designate such a matrix in this section, the size of N being determined in each case by the context.

Let $p = p(x)$ be any polynomial of degree t over $\mathfrak{F}$, $C(p)$ its companion matrix, and $M = C_e(p)$ the matrix

$$C_e(p) = \begin{bmatrix} C(p) & N & 0 & \cdots & 0 \\ 0 & C(p) & N & \cdots & 0 \\ \cdot & \cdot & \cdot & & \cdot \\ \cdot & \cdot & \cdot & & \cdot \\ \cdot & \cdot & \cdot & & \cdot \\ 0 & 0 & 0 & \cdots & N \\ 0 & 0 & 0 & \cdots & C(p) \end{bmatrix},$$

where e is the number of blocks $C(p)$ on the diagonal. Thus M is an $e \times e$ matrix of square blocks, each one $t \times t$, so that M as a matrix of scalars is $n \times n$, $n = et$. The effect of the matrices N is to produce an unbroken line of ones just above the diagonal of M.

LEMMA 8–6. *The matrix $M = C_e(p)$ has minimum and characteristic polynomials both equal to $p(x)^e$. Hence it is indecomposable over $\mathfrak{F}$ if $p(x)$ is irreducible over $\mathfrak{F}$.*

The $n \times n$ matrix $xI - M$ has just above its diagonal an unbroken line of elements equal to -1. This line is the diagonal of the submatrix S obtained by deleting column 1 and row n. All elements above the diagonal of S are zero, so that

$$|S| = (-1)^{n-1} = \pm 1.$$

Then the gcd g_{n-1} of all subdeterminants of $n - 1$ rows and columns is $g_{n-1} = 1$. Since $g_n = |xI - M|$ is the characteristic polynomial of M and $g_n/g_{n-1} = g_n$ is the minimum polynomial (Theorem 8–3), M has the property that its minimum and characteristic polynomials are equal. It remains to show that this polynomial is $p(x)^e$. But

$$|xI - M| = |xI_t - C(p)|^e = p(x)^e.$$

The final statement in the lemma amounts to Theorem 8–8.

If $p = p(x)$ is irreducible over $\mathfrak{F}$, the matrix $M = C_e(p)$ above will be called the *hypercompanion* matrix of $p(x)^e$.

THEOREM 8–10. *Every square matrix is similar to the direct sum of the hypercompanion matrices of its elementary divisors.*

We may start the proof with the canonical matrix of Theorem 8–9, which is the direct sum of the companion matrices $C(p^e)$ of the various elementary divisors $p(x)^e$. It then suffices to show that each matrix $C(p^e)$ is similar to the hypercompanion matrix $C_e(p)$. But this is true by Lemma 8–1, since both matrices have $p(x)^e$ as minimum and characteristic polynomial.

This result furnishes for each A a matrix B which is unique apart from the order in which the hypercompanion matrices appear on the diagonal. If a particular order is agreed upon, Theorem 8–10 gives a canonical set under similarity for matrices over any fixed field $\mathfrak{F}$. It is usual, even without specifying any particular system of order for the diagonal blocks, to refer to any direct sum of hypercompanion matrices as a rational canonical matrix.

The name rational canonical matrix refers (confusingly to the uninitiated) not to the rational number field but to an arbitrary field! Addition, subtraction, multiplication, and division are called rational operations, and these are precisely the operations which can be carried out in any field. For each square matrix A there is a matrix P such that $P^{-1}AP$ is a rational canonical matrix. If all the elements of A lie in a field $\mathfrak{F}$, the matrix P can be chosen to have elements in $\mathfrak{F}$. The name rational canonical matrix* is motivated by the fact that the conversion of A to its canonical matrix $P^{-1}AP$ requires no field larger than the smallest field containing all the elements of A.

The only irreducible polynomials over $\mathfrak{C}$ are linear. Hence each elementary divisor of a complex matrix is a power

$$p(x)^e = (x - a)^e$$

---

* The same argument, however, applies equally well to the canonical matrices of Theorem 8–6, and some writers use the name rational canonical matrix with reference to this theorem.

of a linear polynomial $x - a$, and the corresponding hypercompanion matrix is

(23)
$$C_e(p) = \begin{bmatrix} a & 1 & 0 & \ldots & 0 & 0 \\ 0 & a & 1 & \ldots & 0 & 0 \\ \cdot & \cdot & \cdot & & \cdot & \cdot \\ \cdot & \cdot & \cdot & & \cdot & \cdot \\ \cdot & \cdot & \cdot & & \cdot & \cdot \\ 0 & 0 & 0 & \ldots & a & 1 \\ 0 & 0 & 0 & \ldots & 0 & a \end{bmatrix}.$$

Thus each complex matrix is similar to a direct sum of matrices like (23). This special case of Theorem 8–10 was published by C. Jordan in 1870 and is known variously as the Jordan or *classical canonical* form. Generalizing slightly we obtain:

COROLLARY 8–10. *Let A be a matrix over $\mathfrak{F}$ such that the characteristic polynomial of A factors over $\mathfrak{F}$ into linear factors. Then A is similar to a direct sum of matrices having the form (23), each matrix (23) corresponding to an elementary divisor $(x - a)^e$.*

Suppose that a real matrix A has some elementary divisors which are powers of irreducible quadratics. In this case A is not similar over $\mathfrak{R}$ to a classical canonical matrix. Over $\mathfrak{C}$, however, the same matrix A is similar to such a matrix.

An elementary divisor which is linear is called *simple*. Then (23) becomes the 1×1 matrix (a). The case in which all elementary divisors are simple is of considerable interest because the rational (and classical) canonical matrices are then diagonal.

EXERCISES

1. Let A be an arbitrary square matrix over the rational number field, and consider the three canonical matrices similar to A which are defined in Theorems 8–6, 8–9, and 8–10. Which of these canonical matrices need not remain canonical when all possible enlargements of the scalar field are contemplated?

2. Show that the elementary divisors of A are all linear if and only if the minimum polynomial is a product of distinct linear factors. Is the same true for the characteristic polynomial?

3. The following matrices over the rational number field are in the canonical form of Theorem 8–10. List all the elementary divisors and simi-

larity invariants, and write the corresponding canonical matrices of Theorem 8–6.

(a)
$$\begin{bmatrix} 0 & 1 & 0 & 0 & 0 & 0 & 0 & 0 \\ 2 & 0 & 1 & 0 & 0 & 0 & 0 & 0 \\ 0 & 0 & 0 & 1 & 0 & 0 & 0 & 0 \\ 0 & 0 & 2 & 0 & 1 & 0 & 0 & 0 \\ 0 & 0 & 0 & 0 & 0 & 1 & 0 & 0 \\ 0 & 0 & 0 & 0 & 2 & 0 & 0 & 0 \\ 0 & 0 & 0 & 0 & 0 & 0 & -1 & 0 \\ 0 & 0 & 0 & 0 & 0 & 0 & 0 & -1 \end{bmatrix}.$$

(b)
$$\begin{bmatrix} 2 & 1 & 0 & 0 & 0 \\ 0 & 2 & 1 & 0 & 0 \\ 0 & 0 & 2 & 0 & 0 \\ 0 & 0 & 0 & -1 & 0 \\ 0 & 0 & 0 & 0 & -1 \end{bmatrix}.$$

(c)
$$\begin{bmatrix} 2 & 0 & 0 & 0 & 0 \\ 0 & 2 & 0 & 0 & 0 \\ 0 & 0 & 2 & 0 & 0 \\ 0 & 0 & 0 & -1 & 0 \\ 0 & 0 & 0 & 0 & -1 \end{bmatrix}.$$

4. Each of the following matrices over the rational number field $\mathfrak{R}_0$ is in one or more of the canonical forms of Theorems 8–6, 8–9, 8–10. For each matrix determine which canonical forms already apply and find the remaining canonical forms over $\mathfrak{R}_0$ similar to the given matrix.

(a) The matrix in Exercise 3(a).
(b) The matrix in Exercise 3(c).

(c)
$$\begin{bmatrix} 1 & 0 & 0 \\ 0 & 1 & 0 \\ 0 & 0 & 2 \end{bmatrix}.$$

(d)
$$\begin{bmatrix} 0 & 1 & 0 & 0 \\ 0 & 0 & 1 & 0 \\ 1 & -3 & 3 & 0 \\ 0 & 0 & 0 & 1 \end{bmatrix}.$$

(e)
$$\begin{bmatrix} 0 & 1 & 0 & 0 & 0 & 0 \\ 0 & 0 & 1 & 0 & 0 & 0 \\ 0 & 0 & 0 & 1 & 0 & 0 \\ -4 & 0 & 4 & 0 & 0 & 0 \\ 0 & 0 & 0 & 0 & 0 & 1 \\ 0 & 0 & 0 & 0 & 2 & 0 \end{bmatrix}.$$

(f)
$$\begin{bmatrix} 0 & 1 & 0 & 0 & 0 & 0 \\ 2 & 0 & 1 & 0 & 0 & 0 \\ 0 & 0 & 0 & 1 & 0 & 0 \\ 0 & 0 & 2 & 0 & 0 & 0 \\ 0 & 0 & 0 & 0 & 0 & 1 \\ 0 & 0 & 0 & 0 & 2 & 0 \end{bmatrix}.$$

5. For each matrix of Exercise 4 find all three canonical forms over $\mathfrak{R}$.

6. Find a polynomial $p = p(x)$ such that $C(p^e) = C_e(p)$ for every positive integer e.

7. Find a necessary and sufficient condition on the elementary divisors of A in order that its canonical matrices of Theorem 8–9 and 8–10 may be identical. Hint: Do not overlook elementary divisors of the type x^e.

8. Find a necessary and sufficient condition on the similarity invariants of a matrix A over a field $\mathfrak{F}$ in order that its canonical matrices of Theorems 8–6 and 8–9 may be identical. Then express the condition in terms of the characteristic polynomial. Show that every indecomposable matrix over $\mathfrak{F}$ obeys the condition.

9. Show that every diagonal matrix is in the canonical forms of Theorems 8–9 and 8–10, but is in the form of Theorem 8–6 if and only if it is a scalar matrix.

APPENDIX

8-10 Differential equations and similarity. Consider a system of differential equations

$$(24) \qquad a_{i1}\dot{y}_1 + \cdots + a_{in}\dot{y}_n = b_{i1}y_1 + \cdots + b_{in}y_n,$$

where $i = 1, \ldots, n$, the coefficients a_{ij} and b_{ij} are constants, the y_j are unknown functions of t, and $\dot{y}_j$ means

$$\dot{y}_j = \frac{dy_j}{dt}. \qquad\qquad (j = 1, \ldots, n)$$

This system is much simpler than that considered at the end of Chapter 7, but we shall use it only to demonstrate one place where similarity theory is valuable.

Let

$$A = (a_{ij}), \quad B = (b_{ij}),$$
$$Y = \text{col}\,(y_1, \ldots, y_n), \quad \dot{Y} = \text{col}\,(\dot{y}_1, \ldots, \dot{y}_n)$$

so that the system (24) may be written as

$$(25) \qquad\qquad A\dot{Y} = BY.$$

Very often A is nonsingular, whence if $C = A^{-1}B$, the system reduces to

$$(26) \qquad\qquad \dot{Y} = CY.$$

If we try to simplify the system (26) by a nonsingular linear substitution

$$y_j = \Sigma p_{jk} u_k,$$

where $u_1, \ldots, u_n$ are new unknown functions and $P = (p_{jk})$ is a matrix of constants, we find that

$$\dot{y}_j = \Sigma p_{jk} \dot{u}_k.$$

Letting $U = \mathrm{col}\ (u_1, \ldots, u_n)$, $\dot{U} = \mathrm{col}\ (\dot{u}_1, \ldots, \dot{u}_n)$, we have

$$\begin{aligned}
(27) \qquad & Y = PU, \quad \dot{Y} = P\dot{U} = CPU, \\
& \dot{U} = (P^{-1}CP)U.
\end{aligned}$$

Thus a nonsingular linear substitution on the y_j converts the system (26) with coefficient matrix C into a new system (27) with coefficient matrix $D = P^{-1}CP$, which is similar to C.

This excursion into differential equations points out one way in which similarity arises. It would be particularly beneficial to find, if possible, a similarity

$$P^{-1}CP = D$$

producing a matrix D which is diagonal:

$$D = \mathrm{diag}\ (r_1, \ldots, r_n).$$

In this case the new system (27) has its variables separated,

$$\frac{du_j}{dt} = r_j u_j, \qquad\qquad (j = 1, \ldots, n)$$

and can be integrated at once. The functions y_j are then found by the formula $Y = PU$.

8–11 Characteristic roots of a function of a matrix. The roots of the characteristic polynomial of a square matrix A are called the characteristic roots of A (see Chapter 9). Their importance for the study of A is attested by the classical canonical form (in which the characteristic roots appear on the diagonal) and by the developments in Chapter 9.

Let $r_1, \ldots, r_n$ denote the characteristic roots of A. If $g(x)$ is any polynomial over the scalar field $\mathfrak{F}$, we seek the characteristic roots of $g(A)$. In Exercise 7 at the end of Section 7–9 a method is suggested for proving the interesting and useful fact that the characteristic roots of $g(A)$ are precisely $g(r_1), \ldots, g(r_n)$. We wish to show here how the classical canonical form may be used to give a simple proof of the same result. To accommodate this method we make the scalar field $\mathfrak{F}$ so large that it includes all the characteristic roots, $r_1, \ldots, r_n$, of A. Then A is similar to a matrix C which is a direct sum of matrices like

$$(28) \qquad \begin{bmatrix} r & 1 & 0 & \ldots & 0 & 0 \\ 0 & r & 1 & \ldots & 0 & 0 \\ \cdot & \cdot & \cdot & & \cdot & \cdot \\ \cdot & \cdot & \cdot & & \cdot & \cdot \\ \cdot & \cdot & \cdot & & \cdot & \cdot \\ 0 & 0 & 0 & \ldots & r & 1 \\ 0 & 0 & 0 & \ldots & 0 & r \end{bmatrix} = rI + N,$$

where r is one of the r_i, and the n diagonal elements of C are $r_1, \ldots, r_n$. Also, $g(A)$ is similar to $g(C)$, as brought out in Section 8–2, Exercise 4. Therefore, $g(A)$ and $g(C)$ have the same characteristic roots, so that it suffices to show that the characteristic roots of $g(C)$ are $g(r_1), \ldots, g(r_n)$.

Since C has only zeros below its diagonal, a little computation shows that every power C^k also has this property, and that the diagonal elements of C^k are precisely

$$(29) \qquad r_1^k, \ldots, r_n^k.$$

It follows that $g(C)$ has only zeros below the diagonal, and has $g(r_1), \ldots, g(r_n)$ on its diagonal. Then

$$|xI - g(C)| = [x - g(r_1)] \cdots [x - g(r_n)],$$

so that $g(C)$ has the alleged characteristic roots $g(r_1), \ldots, g(r_n)$.

(The computation of the powers C^k, required in the proof above, may be done conveniently in the following way. First observe that rI commutes with every matrix, in particular with N in (28). But when two matrices, A and B, commute, it may be shown that the binomial formula for powers $(A + B)^k$ is valid. Accepting this fact, we find that

$$(rI + N)^k = r^kI + kr^{k-1}N + \cdots + N^k.$$

Notice also that N^k has ones on the kth slant above the diagonal, and zeros elsewhere, and that $N^k = 0$ if $k \geq t$, where N is $t \times t$. It follows that C^k, which is a direct sum of matrices like $(rI + N)^k$ has zeros everywhere below its diagonal and the quantities (29) on its diagonal. This is the fact needed above.)

It is easy to generalize the above result about the characteristic roots of $g(A)$ to the case in which $g(x)$ is a rational function of x.

THEOREM 8-11. *Let A be a square matrix over $\mathfrak{F}$ and let*

$$g(x) = c(x)/d(x),$$

where $c(x)$ and $d(x)$ are polynomials over $\mathfrak{F}$ and $d(A)$ is nonsingular. Then if $r_1, \ldots, r_n$ are the characteristic roots of A, $g(r_1), \ldots, g(r_n)$ are the characteristic roots of $g(A)$.

There appears to be ambiguity here as to whether $g(A)$ means $c(A)d(A)^{-1}$ or $d(A)^{-1}c(A)$. But, as the proof below shows,*

$$u(A) = d(A)^{-1}$$

is a polynomial in A with scalar coefficients. Hence $u(A)$ commutes with $c(A)$, thus dispelling all ambiguity in the definition of $g(A)$.

The hypothesis that $d(A)$ is nonsingular is equivalent to the condition that $d(x)$ is relatively prime to the characteristic polynomial $f(x)$ of A (see Section 7-9, Exercise 5). Thus

(30) $$u(x)d(x) + v(x)f(x) - 1 = 0$$

for suitable polynomials $u(x)$ and $v(x)$, so that

$$u(A)d(A) - I = 0.$$

Hence $u(A)$ is the inverse of $d(A)$, and if we consider the polynomial

$$h(x) = c(x)u(x),$$

we have

$$h(A) = g(A).$$

Since $h(x)$ is a polynomial, by the proof made earlier we know that $h(r_1), \ldots, h(r_n)$ are the characteristic roots of $h(A) = g(A)$. It remains only to prove that $h(r_i) = g(r_i)$, $i = 1, \ldots, n$. But from (30) and the fact that $f(r_i) = 0$, it follows that

$$u(r_i)d(r_i) = 1, \quad d(r_i)^{-1} = u(r_i),$$

$$g(r_i) = \frac{c(r_i)}{d(r_i)} = c(r_i)u(r_i) = h(r_i).$$

This completes the proof.

---

* We already know that $d(A)^{-1}$ is a polynomial in $d(A)$, hence a polynomial in A.

CHAPTER 9

CHARACTERISTIC ROOTS

9-1 Characteristic vectors. If A is an $n \times n$ matrix over $\mathfrak{F}$ with characteristic function $f(x)$, the scalar field $\mathfrak{F}$ may be taken to be so large that $f(x)$ factors linearly:

$$(1) \qquad f(x) = (x - r_1) \cdots (x - r_n).$$

The roots $r_1, \ldots, r_n$ of $f(x)$ are called the *characteristic roots* of A (also called *latent roots, proper values, secular roots, eigenvalues*). From the definition of the characteristic polynomial it is clear that

$$|r_i I - A| = 0. \qquad\qquad (i = 1, \ldots, n)$$

Although an $n \times n$ matrix always has n characteristic roots, they need not be distinct. The real matrix $A = \mathrm{diag}\,(1, 1, 1, 2)$ has characteristic roots 1, 1, 1, and 2. We say that 1 and 2 are its distinct characteristic roots, and that these roots have multiplicities three and one, respectively. In general, if (1) is the characteristic polynomial of A, the *multiplicity* of a root r is the number of r_i which are equal to r.

In case A (with $r_1, \ldots, r_n$ as its characteristic roots) is similar to a diagonal matrix

$$(2) \qquad P^{-1}AP = D = \mathrm{diag}\,(d_1, \ldots, d_n),$$

the matrices A and D have the same characteristic polynomial $f(x) = (x - r_1) \cdots (x - r_n) = |xI - D| = (x - d_1) \cdots (x - d_n)$. Thus the diagonal elements $d_1, \ldots, d_n$ must be the same as $r_1, \ldots, r_n$ in some order. Hence:

LEMMA 9-1. *If a matrix A is similar to a diagonal matrix D, the characteristic roots of A are the diagonal elements of D.*

By Lemma 8-3 the characteristic roots may be made to appear on the diagonal of D in any desired order by suitable choice of P in (2).

If we write (2) as

$$AP = PD = P\,\mathrm{diag}\,(r_1, \ldots, r_n),$$

and let P_i denote the ith column of P, we find that

$$AP_i = r_i P_i.$$

Thus, multiplication of the vector P_i on the left by A produces a scalar multiple of P_i.

If A is any square matrix and ξ is a *nonzero* column vector such that

$$A\xi = r\xi$$

for some scalar r, ξ is called a *characteristic vector* of A corresponding to r. (Other names: *proper vector, eigenvector*.)

Clearly ξ is any nontrivial solution of the homogeneous system of linear equations, $(A - rI)\xi = 0$, or equivalently, $(rI - A)\xi = 0$. Such a system (Corollary 3–6) has a nontrivial solution if and only if $|rI - A| = 0$, thus if and only if r is a characteristic root of A. Hence:

THEOREM 9–1. *Let A be a square matrix and r a scalar. Then A has a characteristic vector corresponding to r if and only if r is a characteristic root of A. In this case, moreover, the totality of characteristic vectors of A corresponding to r is the set of all nonzero vectors in the null space of $rI - A$.*

Because of the final property in this theorem, the null space of $rI - A$ may be called the characteristic vector space (eigenvector space) of A corresponding to r.

This chapter is concerned with matrices which are similar to diagonal matrices. The discussion above is a first indication that characteristic roots and vectors may play a prominent role in the theory. We shall adhere to the terminology "null space of $rI - A$," but the reader should remember that, apart from the zero vector, this space consists of the characteristic vectors of A corresponding to the root r.

Throughout this chapter it will be understood that vectors are written as columns unless the contrary is indicated.

EXERCISES

1. For each of the following matrices over $\Re$ find the characteristic roots, and for each root r find all characteristic vectors corresponding to r:

$$\begin{bmatrix} 0 & 1 \\ -2 & 3 \end{bmatrix}, \quad \begin{bmatrix} 2 & -1 \\ 1 & 0 \end{bmatrix}.$$

2. Prove that a square matrix A is nonsingular if and only if all characteristic roots of A are nonzero (a) by use of determinants; (b) by use of Theorem 3–18.

9–2 Similarity to a diagonal matrix. Several criteria for a matrix to be similar to a diagonal matrix will now be developed.

THEOREM 9-2. *An $n \times n$ matrix A over $\mathfrak{F}$ is similar to a diagonal matrix if and only if the set of characteristic vectors of A includes a basis for $V_n(\mathfrak{F})$.*

Suppose that $\xi_1, \ldots, \xi_n$ are characteristic vectors of A forming a basis for $V_n(\mathfrak{F})$, so that

$$(3) \qquad\qquad A\xi_i = r_i\xi_i = \xi_i r_i. \qquad\qquad (i = 1, \ldots, n)$$

If P is the matrix whose ith column is ξ_i $(i = 1, \ldots, n)$ and if D denotes the matrix diag $(r_1, \ldots, r_n)$, equation (3) is merely the column-by-column statement of the equality

$$(4) \qquad\qquad AP = PD.$$

That P is nonsingular follows from the linear independence of the ξ_i. Hence

$$(5) \qquad\qquad P^{-1}AP = D.$$

Conversely, (5) leads through (4) to (3), so that the columns of P are characteristic vectors of A, which because of their linear independence form a basis of $V_n(\mathfrak{F})$. This completes the proof of Theorem 9-2.

Another criterion of the similarity of A to a diagonal matrix is given in Exercise 8 of Section 8-8: all elementary divisors of A must be linear. This can easily be seen to be equivalent to the following criterion:

THEOREM 9-3. *A square matrix A is similar to a diagonal matrix if and only if the minimum polynomial of A factors into a product of distinct linear factors.*

We shall give another proof of this theorem. If the minimum polynomial has the property stated, all of the elementary divisors are linear, and Theorem 8-9 gives a diagonal matrix similar to A. Conversely, if A is similar to $D = $ diag $(r_1, \ldots, r_n)$, A and D have the same minimum polynomial $m(x)$, and it suffices to discuss D. Let

$$(6) \qquad\qquad s_1, \ldots, s_t$$

denote the distinct quantities in the list $r_1, \ldots, r_n$ of characteristic roots of D. Then each r_i is equal to one and only one s_j. The polynomial

$$(7) \qquad\qquad p(x) = (x - s_1) \cdots (x - s_t)$$

has the factorization property described in Theorem 9-3 and the property that $p(r_i) = 0$, $i = 1, \ldots, n$.

If, for example, the r_i are 1, 1, 1, −1, 3, 3, then

$$p(x) = (x - 1)(x + 1)(x - 3).$$

In general, since every $p(r_i) = 0$, we have

$$p(D) = \text{diag } [p(r_1), \ldots, p(r_n)] = 0,$$

whence (Theorem 8–2) $m(x)$ divides $p(x)$. The very definition of $p(x)$ implies that it divides $m(x)$. It follows that $m(x) = p(x)$ and the latter by construction is a product of distinct, linear factors.

This gives a construction for the minimum polynomial of A in terms of the characteristic roots, provided it is known that A is similar to a diagonal matrix.

LEMMA 9–2. *Let $\eta_1, \ldots, \eta_t$ be characteristic vectors of A corresponding to distinct characteristic roots $s_1, \ldots, s_t$, respectively. Then $\eta_1, \ldots, \eta_t$ are linearly independent.*

Suppose that

$$a_1\eta_1 + \cdots + a_t\eta_t = 0$$

for certain scalars a_i. Repeated multiplication on the left by A then gives

$$\begin{aligned}
a_1\eta_1 \quad &+ \cdots + a_t\eta_t \quad = 0, \\
a_1\eta_1 s_1 \quad &+ \cdots + a_t\eta_t s_t \quad = 0, \\
a_1\eta_1 s_1^2 \quad &+ \cdots + a_t\eta_t s_t^2 \quad = 0, \\
& \cdot \\
& \cdot \\
& \cdot \\
a_1\eta_1 s_1^{t-1} &+ \cdots + a_t\eta_t s_t^{t-1} = 0.
\end{aligned}$$

This system of t equations may be written in the following compact fashion in which the vectors $a_i\eta_i$ appear as columns of a larger matrix:

(8) $$(a_1\eta_1, \ldots, a_t\eta_t)S = 0,$$

$$S = \begin{bmatrix} 1 & s_1 & s_1^2 & \ldots & s_1^{t-1} \\ \cdot & \cdot & \cdot & & \cdot \\ \cdot & \cdot & \cdot & & \cdot \\ \cdot & \cdot & \cdot & & \cdot \\ 1 & s_t & s_t^2 & \ldots & s_t^{t-1} \end{bmatrix}.$$

Since S is a Vandermonde matrix with distinct s_i, it is nonsingular (Exercise 8, Section 4–7). Multiplication of (8) on the right by S^{-1} gives

$$(a_1\eta_1, \ldots, a_t\eta_t) = 0,$$

whence each column $a_i\eta_i = 0$. The characteristic vectors η_i are not zero, so every a_i must be zero. This establishes the independence of the η_i.

THEOREM 9-4. *A square matrix A is similar to a diagonal matrix if and only if for each characteristic root r of A, the multiplicity of r equals the dimension of the null space of* $rI - A$.

Before proceeding to the proof we may reinterpret the latter property: if A is similar to a diagonal matrix, the multiplicity of r equals the maximum number of linearly independent characteristic vectors of A corresponding to r.

For proof let $P^{-1}AP = D = \text{diag} (r_1, \ldots, r_n)$, where exactly k of the r_i are equal to r. Then $rI - D$ has precisely k zeros on the diagonal, hence has rank $n - k$, whence (Theorem 3-9) the null space of $rI - D$ has dimension $n - (n - k) = k$. Since

$$rI - A = P(rI - D)P^{-1},$$

$rI - A$ has the same rank $n - k$ and nullity k as $rI - D$. To prove the converse, let $s_1, \ldots, s_t$ denote the distinct characteristic roots of A, with m_i both as the multiplicity of s_i and the dimension of the null space N_i of $s_iI - A$ $(i = 1, \ldots, t)$. Then $n = m_1 + \cdots + m_t$, where A is $n \times n$. If we choose a basis

$$\text{(9)} \qquad \xi_{i1}, \ldots, \xi_{im_i}$$

of the space N_i $(i = 1, \ldots, t)$, all of the ξ_{ij} form a set of n vectors which, as we shall show, constitute a basis for $V_n(\mathfrak{F})$. If

$$\sum_{i,j} a_{ij}\xi_{ij} = 0,$$

we may first combine those terms belonging to a common null space N_i to obtain an equation of the form

$$\text{(10)} \qquad \eta_1 + \cdots + \eta_t = 0, \quad \eta_i = \sum_j a_{ij}\xi_{ij},$$

where each η_i belongs to N_i. Hence each η_i is either 0 or a characteristic vector of A corresponding to s_i. The latter property is untenable, since then Lemma 9-2 would assert the independence of the η_i, contrary to (10). Thus each

$$\eta_i = 0 = \sum_j a_{ij}\xi_{ij},$$

and the independence of the basis vectors (9) implies that every $a_{ij} = 0$. This proves the independence of the set of n vectors ξ_{ij}, which thus form a basis of $V_n(\mathfrak{F})$ consisting of characteristic vectors of A. Then A is similar to a diagonal matrix by virtue of Theorem 9–2.

EXERCISES

1. Show how to prove Theorem 9–3 by means of Exercise 8, Section 8–8.

2. Find the characteristic roots of each of the following matrices over $\mathfrak{R}$ and for each root a characteristic vector. Follow the proof of Theorem 9–2 to construct a matrix P such that $P^{-1}AP$ is diagonal.

$$A = \begin{bmatrix} 1 & 2 \\ 2 & 1 \end{bmatrix}; \quad A = \begin{bmatrix} 1 & 2 \\ 3 & 0 \end{bmatrix}.$$

3. Find the minimum polynomial of each of the matrices in Exercise 2, and verify Theorem 9–3 in these instances.

4. Let A be similar to a diagonal matrix and let $g(x)$ be any polynomial. Prove that $g(A)$ is similar to a diagonal matrix.

5. If the characteristic polynomial of A is a product of distinct, linear factors, prove that A is similar to a diagonal matrix. Is the converse true?

6. If A is idempotent and has rank r, prove without using the canonical matrices of Chapter 8 that A is similar to the diagonal matrix

$$\begin{bmatrix} I_r & 0 \\ 0 & 0 \end{bmatrix}.$$

Hence prove that two $n \times n$ idempotent matrices over the same field are similar if and only if they have the same rank.

7. (a) Use the proof of Theorem 9–3 to formulate a construction for the minimum polynomial of A from its list of characteristic roots, provided it is known that A is similar to a diagonal matrix. (b) Give an example of a matrix with characteristic roots 1, 2, 2 whose minimum polynomial is not $(x-1)(x-2)$.

8. Let A be *triangular*, that is, a square matrix in which every element above the diagonal is zero. Prove that A is similar to a diagonal matrix if the diagonal elements of A are distinct.

9. If A is similar to a diagonal matrix, prove that its transpose also has this property.

9–3 Decomposition into principal idempotents.
Diagonal matrices are very easy to manipulate algebraically. Perhaps the best support for this assertion lies in the fact, already used several times, that if $p(x)$ is any polynomial and if $D = \text{diag}(r_1, \ldots, r_n)$, then

$$p(D) = \text{diag}[p(r_1), \ldots, p(r_n)].$$

A matrix A which is similar to a diagonal matrix enjoys much of this simplicity when A is viewed appropriately. The next result sets forth this viewpoint.

THEOREM 9-5. *Let A be an $n \times n$ matrix whose distinct characteristic roots are $s_1, \ldots, s_t$. Then, A is similar to a diagonal matrix if and only if there exist matrices $E_1, \ldots, E_t$ such that:*

(a) $A = s_1 E_1 + \cdots + s_t E_t$;

(b) *each E_i is idempotent* $(E_i^2 = E_i)$;

(c) $E_i E_j = 0$ *if* $i \neq j$;

(d) $E_1 + \cdots + E_t = I$.

Moreover,

(e) *the matrices E_i are uniquely determined by A and properties* (a), (b), (c), *and* (d);

(f) *each E_i has rank equal to the multiplicity of the characteristic root s_i;*

(g) *if $p(x)$ is any polynomial, $p(A) = p(s_1)E_1 + \cdots + p(s_t)E_t$;*

(h) *any given matrix B commutes with A if and only if B commutes with every E_i.*

If m_i denotes the multiplicity of s_i, A is similar to

$$(11) \qquad P^{-1}AP = \operatorname{diag}\,(s_1 I_{m_1}, \ldots, s_t I_{m_t}).$$

To put this equation in a form more convenient for present purposes, let D_i denote an $n \times n$ diagonal matrix obtainable from the right side of (11) by use of 1 in place of s_i and 0 in place of s_j for every $j \neq i$. Then

$$I = D_1 + \cdots + D_t,$$
$$P^{-1}AP = s_1 D_1 + \cdots + s_t D_t,$$
$$A = s_1 P D_1 P^{-1} + \cdots + s_t P D_t P^{-1}.$$

If we define $E_i = P D_i P^{-1}$ $(i = 1, \ldots, t)$, we have (a). Moreover, since $D_i^2 = D_i$ and $D_i D_j = 0$ when $i \neq j$, it follows that

$$E_i^2 = P D_i P^{-1} P D_i P^{-1} = P D_i P^{-1} = E_i,$$
$$E_i E_j = P D_i D_j P^{-1} = 0, \qquad (i \neq j)$$
$$E_1 + \cdots + E_t = P(D_1 + \cdots + D_t)P^{-1} = P I P^{-1} = I.$$

Thus (b), (c), and (d) are established.

Before proceeding to the converse, we note that properties (f) and (g) may be easily proved. Since D_i has rank m_i, the same is true for the similar matrix E_i, which is the assertion in (f). To get property (g) the reader may use (a), (b), and (c) to make computations like the following:

$$A^2 = (s_1 E_1 + \cdots + s_t E_t)(s_1 E_1 + \cdots + s_t E_t)$$
$$= \sum_{i,j} s_i E_i s_j E_j = \sum_i s_i^2 E_i^2 = \sum_i s_i^2 E_i$$

and, more generally,

$$(12) \qquad A^k = \sum_i s_i^k E_i.$$

If

$$p(x) = a_r x^r + \cdots + a_1 x + a_0,$$

we form $p(A)$ by multiplying (12) by a_k ($k = 1, \ldots, r$) and adding the resulting r equations together with the equation

$$a_0 I = a_0 E_1 + \cdots + a_0 E_t.$$

This gives property (g).

To prove the converse part of Theorem 9–5 it suffices to show that the minimum polynomial $m(x)$ of A factors into a product of distinct, linear factors. This in turn will be true if we find a polynomial $p(x)$ which factors into distinct, linear factors and has the property that $p(A) = 0$. (Why does this suffice?) Define

$$p(x) = (x - s_1) \cdots (x - s_t).$$

Since the s_i are distinct, it remains only to show that $p(A) = 0$. But every $p(s_i) = 0$ and

$$p(A) = p(s_1)E_1 + \cdots + p(s_t)E_t = 0.$$

The proof of the converse is complete.

It remains only to prove (e) and (h). For (e) let

$$A = s_1 F_1 + \cdots + s_t F_t,$$

where $F_1 + \cdots + F_t = I$, $F_i^2 = F_i$, and $F_i F_j = 0$ if $i \neq j$. First note that

$$E_i A = A E_i = s_i E_i, \quad F_j A = A F_j = s_j F_j.$$

Thus

$$E_i(A F_j) = E_i s_j F_j = (E_i A) F_j = s_i E_i F_j,$$
$$(s_j - s_i) E_i F_j = 0,$$
$$E_i F_j = 0. \qquad (i \neq j)$$

This fact permits the following computation:

$$E_i = E_i I = E_i(\Sigma F_j) = E_i F_i$$
$$= (\Sigma E_j) F_i = I F_i = F_i.$$

The uniqueness property (e) is thus established. The matrices $E_1, \ldots, E_t$, uniquely determined by A, are called the *principal idempotents* of A.

It remains to prove (h). If B commutes with each E_i, it certainly commutes with A by property (a). If, conversely, B commutes with A it must commute with every polynomial in A. The proof of (h) will then be complete if we show that each E_i is in fact a polynomial in A.

Let

$$p(x) = (x - s_1) \cdots (x - s_t)$$

and let

$$p_i(x) = p(x)/(x - s_i).$$

Thus $p_i(x)$ is a polynomial of degree $t - 1$ whose roots are those s_j with $j \neq i$. Then

$$p_i(A) = p_i(s_1)E_1 + \cdots + p_i(s_t)E_t$$
$$= p_i(s_i)E_i.$$

Since

$$c_i = p_i(s_i)$$

is a nonzero scalar, we may define the polynomial

$$q_i(x) = c_i^{-1}p_i(x)$$

which has the property $q_i(A) = E_i$. This demonstration that each E_i is a polynomial in A completes the proof.

In certain studies the set of all characteristic roots of A is known as the "spectrum" of A, and formula (a) in Theorem 9–5 is then called the "spectral decomposition" of A. The same result is also known as the "decomposition of A into principal idempotents."

We now have a variety of criteria for a matrix to be similar to a diagonal matrix (such matrices are called *diagonable*), or, viewed differently, we have a variety of properties of matrices which are diagonable. In the sections ahead we shall study specialized types of similarity which are of importance for real and complex matrices. Several routes are available for the journey through these theories. The route to be followed here is marked out in large part by Theorems 9–2 and 9–4.

EXERCISES

1. Answer the question raised in the proof of the converse part of Theorem 9–5.

2. Find the decomposition into principal idempotents for the matrix

$$A = \begin{bmatrix} 1 & 2 \\ 3 & 0 \end{bmatrix},$$

and verify properties (b), (c), and (d) of Theorem 9–5 for this matrix.

9–4 Inner products. The next few sections concern orthogonal similarity (defined in Section 9–5) and lead to a famous result

(Theorem 9–11) about real, symmetric matrices. In this section the underlying concepts will be introduced.

The field $\Re$ of all real numbers will be the field of scalars now. If ξ and η belong to $V_n(\Re)$ a particular scalar will be associated with them. Let

$$\xi = \mathrm{col}\ (a_1, \ldots, a_n), \quad \eta = \mathrm{col}\ (b_1, \ldots, b_n).$$

Then the scalar

(13) $$\xi'\eta = a_1 b_1 + \cdots + a_n b_n$$

is called the *inner product* of ξ and η. The inner product of ξ with itself is a sum of squares:

$$\xi'\xi = a_1^2 + \cdots + a_n^2.$$

It is here that the use of the real numbers as scalars begins to have influence, for a sum of squares of real numbers a_i is never zero unless every $a_i = 0$.

LEMMA 9–3. *The inner product of a vector of $V_n(\Re)$ with itself is positive unless the vector is zero.*

Since nonnegative real numbers have square roots in the real number system, we may speak of the square root

(14) $$(\xi'\xi)^{\frac{1}{2}},$$

which is taken always to be nonnegative.

DEFINITION 1. If ξ is any vector of $V_n(\Re)$ the *length* of ξ is defined by the formula (14) as the square root of the inner product of ξ and ξ. A vector is called *normal* if its length is unity.

DEFINITION 2. Two vectors of $V_n(\Re)$ are said to be *orthogonal* to each other if their inner product is zero.

These concepts of length and orthogonality have the usual connotations when $n = 2$ or 3. Suppose that the vector (a_1, a_2) of $V_2(\Re)$ is interpreted as the line from the origin O to the point P, whose coordinates in some rectangular coordinate system are a_1 and a_2. Then the length of OP is $(a_1^2 + a_2^2)^{\frac{1}{2}}$. Moreover, if Q is the point (b_1, b_2), OQ has slope b_2/b_1 and OP has slope a_2/a_1, on the assumption that a_1 and b_1 are not zero. The familiar criterion for orthogonality of OQ and OP is that the product of their slopes shall be -1:

$$\frac{b_2}{b_1}\frac{a_2}{a_1} = -1.$$

This condition may be stated in the equivalent form $b_2a_2 = -b_1a_1$, or as

$$a_1b_1 + a_2b_2 = 0.$$

Thus for $V_2(\mathfrak{R})$ orthogonality in the usual geometric sense is the same as orthogonality in the sense of Definition 2. The same interpretation can be made for $V_3(\mathfrak{R})$.

Let $P = (p_{ij})$ be an $n \times n$ matrix and let $\xi = \text{col } (x_1, \ldots, x_n)$. Then $P\xi = \eta$ is a vector, $\text{col } (y_1, \ldots, y_n)$, and the equation

(15) $$\eta = P\xi$$

is called a *linear transformation* of $V_n(\mathfrak{R})$. Such transformations are studied in some detail in the next chapter. For the present it should be noted that, given P, the formula (15) associates with each vector ξ of $V_n(\mathfrak{R})$ a well-defined vector η which is variously called the *image* or *map* or *transform* of ξ.

DEFINITION 3. A matrix P and the linear transformation (15) which P determines are both called *orthogonal* if (15) preserves lengths, that is,

$$(\xi'\xi)^{\frac{1}{2}} = (\eta'\eta)^{\frac{1}{2}}$$

for every vector ξ of $V_n(\mathfrak{R})$.

It is easy to characterize orthogonal matrices $P = (p_{ij})$ in terms of their elements. For if P is orthogonal,

(16) $$\xi'\xi = \eta'\eta = (P\xi)'P\xi = \xi'P'P\xi.$$

Since this asserts that the quadratic forms

$$X'IX, \quad X'(P'P)X$$

have the same values, Theorem 5–3 implies that

(17) $$P'P = I,$$

which is equivalent to the statement, $P' = P^{-1}$. Conversely, (17) implies (16), so that (15) is length-preserving.

THEOREM 9–6. *A square matrix P is orthogonal if and only if $P' = P^{-1}$. This is true, also, if and only if the columns P_i of P are mutually orthogonal, normal vectors:*

$$P_i'P_i = 1, \quad P_i'P_j = 0. \qquad (i \neq j)$$

The first criterion in Theorem 9–6 was proved above. The second is merely a row-by-column interpretation of equation (17).

Although a matrix P over an arbitrary field $\mathfrak{F}$ may be called orthog-onal if it obeys (17), the bulk of the literature on orthogonal matrices is confined to the case $\mathfrak{F} = \mathfrak{R}$. We shall abide by the definition above, whereby it is always assumed that the scalars are real when a matrix is called orthogonal.

DEFINITION 4. A basis for a vector space over $\mathfrak{R}$ is called *ortho-normal* if it consists of vectors which are normal and mutually orthog-onal.

Since an $n \times n$ orthogonal matrix is nonsingular, its columns form a basis for $V_n(\mathfrak{R})$. By Theorem 9–6 this basis is orthonormal. The unit vectors constitute an orthonormal basis.

EXERCISES

1. Prove that a matrix is orthogonal if and only if its rows are mutually orthogonal, normal vectors.

2. Show that every orthogonal matrix commutes with its transpose.

★3. Prove that a product of orthogonal matrices is always orthogonal.

★4. Prove that the inverse of an orthogonal matrix is orthogonal.

★5. If ξ and η are orthogonal to each other, show that the same is true of $a\xi$ and $b\eta$ for any scalars a and b.

★6. Let ξ be a nonzero vector of length c. Show that the vector $c^{-1}\xi$ is normal.

9–5 Orthogonal similarity. If $B = P^{-1}AP$, where P is orthogonal, B is said to be *orthogonally similar* to A (also: orthogonally congruent or equivalent). Then B is simultaneously similar and congruent to A, the same matrix P functioning in both relations:

$$B = P^{-1}AP = P'AP, \quad P^{-1} = P'.$$

This relation of orthogonal similarity is clearly reflexive ($A = I^{-1}AI$, $I^{-1} = I'$); and since a matrix is orthogonal if it is the inverse of an orthogonal matrix or the product of two orthogonal matrices, the relation is also symmetric and transitive.

If A is symmetric, every matrix orthogonally similar to A is sym-metric, since orthogonal similarity is also congruence. We shall ob-tain a canonical set (Theorem 9–11) for real, symmetric matrices under the relation of orthogonal similarity. Since the matrices in this canonical set will be diagonal, we now embark on a preliminary ex-ploration, as in Theorem 9–2, of those matrices which are orthogonally similar to diagonal matrices:

(18) $P^{-1}AP = D = \text{diag}(r_1, \ldots, r_n), \quad P^{-1} = P'.$

Suppose that (18) holds, so that $AP = PD$. If P_i denotes the ith column of P, then

$$AP_i = r_i P_i,$$

and $P_1, \ldots, P_n$ are characteristic vectors of A. As observed after Definition 4, these vectors form an orthonormal basis for $V_n(\mathfrak{R})$. Thus the validity of (18) implies that the characteristic vectors of A include an orthonormal basis for $V_n(\mathfrak{R})$.

THEOREM 9-7. *Let A be an $n \times n$ matrix over $\mathfrak{R}$. Then A is orthogonally similar to a diagonal matrix if and only if the set of characteristic vectors of A includes an orthonormal basis for $V_n(\mathfrak{R})$.*

It remains to prove the converse. Suppose that $\xi_1, \ldots, \xi_n$ are characteristic vectors of A forming an orthonormal basis for $V_n(\mathfrak{R})$:

$$A\xi_i = r_i \xi_i. \qquad\qquad (i = 1, \ldots, n)$$

This implies that $AP = PD$, $D = \text{diag}(r_1, \ldots, r_n)$, where P is the matrix whose ith column is ξ_i $(i = 1, \ldots, n)$. Since the ξ_i are mutually orthogonal normal vectors, P is orthogonal (Theorem 9-6), so that

$$P^{-1}AP = D, \quad P^{-1} = P'.$$

EXERCISES

★1. Show in detail that orthogonal similarity is an RST relation.

2. Let $B = P^{-1}AP$, where P is orthogonal. Show that all orthogonal matrices Q satisfying $Q^{-1}AQ = B$ are given by $Q = RP$, R varying over all orthogonal matrices commuting with A.

3. Is a nonsingular, real matrix P necessarily orthogonal if it obeys $P^{-1}AP = P'AP$ for some matrix A?

★4. If the columns of an orthogonal matrix are permuted in any way, show that the resulting matrix is orthogonal.

5. Show that the diagonal elements r_i in the proof of Theorem 9-7 may be made to appear on the diagonal in any desired order.

9-6 Orthonormal bases. The unit vectors of $V_n(\mathfrak{R})$ form an orthonormal basis. We shall prove that all nonzero subspaces of $V_n(\mathfrak{R})$ have orthonormal bases.

Since the coordinates of the unit vectors of $V_n(\mathfrak{R})$ lie in the rational number field $\mathfrak{R}_0$, these vectors may be called an orthonormal basis for $V_n(\mathfrak{R}_0)$. Subspaces of $V_n(\mathfrak{R}_0)$, however, do not always have bases of this type. Consider, for example, the subspace S spanned by $\xi = (1, 1, 0, \ldots, 0)$. An arbitrary vector in S has the form $\gamma =$

$(c, c, 0, \ldots, 0)$, and the square of the length of γ is $2c^2$, where c is rational. If this were equal to 1 we should have

$$2c^2 = 1, \quad 2 = (1/c)^2,$$

and $1/c$ would be a rational square root of 2. Since this is impossible, S has no normal vectors, and hence no orthonormal basis. Over the real number field the situation is better, as we shall see.

LEMMA 9–4. *If $\xi_1, \ldots, \xi_t$ are mutually orthogonal nonzero vectors of $V_n(\mathfrak{R})$, they are linearly independent.*

Suppose that

$$a_1\xi_1 + \cdots + a_t\xi_t = 0,$$

with suitable scalars a_i. Since $\xi_i'\xi_j = 0$ for $i \neq j$,

$$\begin{aligned}
0 = \xi_i' \cdot 0 &= \xi_i'(a_1\xi_1 + \cdots + a_t\xi_t) \\
&= a_1(\xi_i'\xi_1) + \cdots + a_t(\xi_i'\xi_t) \\
&= a_i\xi_i'\xi_i.
\end{aligned}$$

Since ξ_i is assumed to be nonzero, $\xi_i'\xi_i$ is nonzero. Therefore $a_i = 0$ $(i = 1, \ldots, t)$ and the set of ξ_i is linearly independent.

THEOREM 9–8. *Every nonzero vector space V over the real number field has an orthonormal basis. Moreover, every set of mutually orthogonal, normal vectors of V may be extended to an orthonormal basis of V.*

Let us take the second statement first, and let

(19) $\xi_1, \ldots, \xi_i$ $(i \geq 1)$

be a set of mutually orthogonal, normal vectors belonging to V. (In case $i = 1$, this merely requires ξ_1 to be a vector of length one.) By the lemma above, the number i is at most equal to the dimension t of V, and the only case in need of consideration is that in which $i < t$.

Since $i < t$, the subspace spanned by (19) is not all of V, so that there is a vector η not in this subspace. Therefore no vector ξ of the form

(20) $\xi = \eta - a_1\xi_1 - \cdots - a_i\xi_i$

can be zero. In (20) choose the scalars a_j to be

(21) $a_j = \xi_j'\eta$

for $j = 1, \ldots, i$. Then (20), when multiplied on the left by ξ_j', becomes

$$\xi_j'\xi = \xi_j'\eta - \sum_{h=1}^{i} a_h\xi_j'\xi_h$$
$$= \xi_j'\eta - a_j\xi_j'\xi_j$$
$$= \xi_j'\eta - a_j = 0.$$

Thus all of the vectors (19) are orthogonal to ξ and also to

$$\xi_{i+1} = c^{-1}\xi,$$

where c is the length of ξ. The set (19) has thus been extended to a set

$$\xi_1, \ldots, \xi_i, \xi_{i+1}$$

of mutually orthogonal, normal vectors of V. If $i + 1 < t$, the process may be repeated, and the desired basis will be constructed after finitely many repetitions. To prove the first part of the theorem we note that every nonzero vector space V possesses a vector $\xi \neq 0$, hence possesses a vector ξ_1 which is normal. This gives a set (19) to which the process above, or the second statement in the theorem, may be applied.

EXERCISES

1. Let V be the subspace of $V_3(\Re)$ spanned by $\xi = (1, 1, 1)$ and $\eta = (1, 2, 3)$. Find two orthonormal bases for V, using the procedure in the proofs above. Then extend each basis to an orthonormal basis for $V_3(\Re)$.

2. Let the columns of an $n \times r$ matrix A form an orthonormal basis for a subspace V of $V_n(\Re)$. If B is $n \times r$, prove that the columns of B form an orthonormal basis of V if and only if $B = AC$, where C is an $r \times r$ orthogonal matrix.

9-7 Similarity of real, symmetric matrices. Most of the remaining tools needed for studying the orthogonal similarity of real, symmetric matrices are developed in this section. Some of the results are extended to the larger class of matrices which are Hermitian, and will be used later when this class is studied relative to a suitable RST relation.

THEOREM 9-9. *All characteristic roots of a Hermitian matrix are real.*

Let A be the matrix and let r denote any one of its characteristic roots, r being a complex number. There is then a nonzero vector ξ over $\mathbb{C}$ such that

$$A\xi = r\xi, \quad \xi = \text{col } (c_1, \ldots, c_n) \neq 0.$$

If, as in Section 5–7, ξ denotes the vector

$$\bar{\xi} = \text{col } (\bar{c}_1, \ldots, \bar{c}_n),$$

we may multiply the equation above by $\bar{\xi}'$ to obtain

$$\bar{\xi}'A\xi = r\bar{\xi}'\xi.$$

The left side of this equation is a value of a Hermitian form, hence by Theorem 5–13 is a real number s.   For the same reason the expression $\bar{\xi}'\xi$ is a real number t, and t cannot be zero:

$$t = \bar{c}_1 c_1 + \cdots + \bar{c}_n c_n > 0.$$

Thus we have $s = rt$, so that r is the real number $r = s/t$.

LEMMA 9–5.   *Let r_1 and r_2 be distinct characteristic roots of a matrix A, and let ξ_i be a characteristic vector of A corresponding to r_i ($i = 1, 2$). Then if A is Hermitian, $\bar{\xi}'_1 \xi_2 = 0$.*

The proof is an application of the associative law and the fact, just proved, that r_1 and r_2 are real:

$$\bar{\xi}'_1(A\xi_2) = \bar{\xi}'_1(r_2\xi_2) = r_2(\bar{\xi}'_1\xi_2)$$
$$= \overline{\bar{\xi}'_1 A}'\xi_2 = (\overline{A\xi_1})'\xi_2$$
$$= (\overline{r_1\xi_1})'\xi_2 = r_1(\bar{\xi}'_1\xi_2).$$

Thus

$$(r_2 - r_1)(\bar{\xi}'_1\xi_2) = 0, \quad \bar{\xi}'_1\xi_2 = 0,$$

since by assumption $r_2 - r_1$ is not zero.

This lemma may be applied to real, symmetric matrices A.   Since we saw in Theorem 9–9 that all characteristic roots of A are real, we may take $\mathcal{R}$ as the scalar field so that ξ_1 and ξ_2 are real.   Then, as applied to this case, Lemma 9–5 may be stated thus: *Characteristic vectors corresponding to distinct roots of a real, symmetric matrix are always orthogonal.*

As a result of Theorem 9–9 the minimum polynomial, in fact every similarity invariant, of a real, symmetric matrix A factors over $\mathcal{R}$ into linear factors.   We shall see that the factors of the minimum polynomial $m(x)$ are distinct.   This is equivalent (Theorem 9–3) to the property that A is similar to a diagonal matrix.

LEMMA 9–6.   *Every real, symmetric matrix is similar to a diagonal matrix.*

Using the notations A and $m(x)$ above, we need only show that the minimum polynomial $m(x)$ has no repeated linear factors.   Suppose to the contrary that

(22) $$m(x) = (x - r)^2 h(x).$$

Then $h(x)$ and $g(x) = (x - r)h(x)$ have real coefficients, and

$$g(x)^2 = m(x)h(x), \quad g(A)^2 = m(A)h(A) = 0.$$

The matrix

$$S = g(A) = (s_{ij})$$

is real and symmetric (why?) and is not zero, since $g(x)$ has lower degree than the minimum polynomial of A. The ith diagonal element of $S^2 = g(A)^2 = 0$ is

$$(s_{i1}, \ldots, s_{in}) \cdot \text{col } (s_{1i}, \ldots, s_{ni}) = 0,$$

and since $s_{ij} = s_{ji}$, this element is

$$s_{i1}^2 + \cdots + s_{in}^2 = 0.$$

The reality of the s_{ij} then implies that every $s_{ij} = 0$, so that $S = 0$, contrary to fact. Thus (22) is false and the proof is complete.

THEOREM 9–10. *Two real, symmetric matrices are similar if and only if they have the same characteristic roots.*

For proof we use the diagonal matrix D, which by Lemma 9–6 is similar to the given matrix A. As observed in Lemma 9–1, the diagonal elements of D are the characteristic roots $r_1, \ldots, r_n$ of A, and these may be made to appear on the diagonal of D in any desired order. If B is a real, symmetric matrix having the same roots as A, it is similar to the same diagonal matrix D, hence to A. The converse is trivial.

EXERCISES

1. Show that the elementary divisors of a real, symmetric matrix are linear.

2. By Theorem 9–10 determine which two of the following real matrices are similar, and write a diagonal matrix similar to both:

$$\begin{bmatrix} -2/5 & 6/5 \\ 6/5 & 7/5 \end{bmatrix}, \quad \begin{bmatrix} 5/2 & 5/2 \\ 5/2 & 1/2 \end{bmatrix}, \quad \begin{bmatrix} 1/2 & 3/2 \\ 3/2 & 1/2 \end{bmatrix}.$$

3. Show that the rank of a real, symmetric matrix A is the number of nonzero characteristic roots of A. Find a larger class of matrices to which this result can be extended. Find a matrix whose rank is not equal to its number of nonzero characteristic roots.

9–8 Orthogonal similarity of symmetric matrices. A canonical set is obtained now for symmetric matrices under orthogonal similarity.

THEOREM 9–11. *Every real, symmetric matrix A is orthogonally similar to a diagonal matrix (whose diagonal elements are necessarily the characteristic roots of A).*

Let $s_1, \ldots, s_t$ denote the distinct characteristic roots of A, and let s_i have multiplicity m_i $(i = 1, \ldots, t)$ so

$$n = m_1 + \cdots + m_t.$$

Since (Lemma 9–6) A is similar to a diagonal matrix, Theorem 9–4 comes into play: m_i is the dimension of the null space N_i of $s_i I - A$. If we choose an orthonormal basis for each N_i these basis vectors form a set of n vectors $\xi_1, \ldots, \xi_n$ which may be seen to be an orthonormal basis for $V_n(\Re)$. For, they are normal and two of them belonging to the same N_i are orthogonal. If two of them belong to N_i and N_j, respectively, with $i \neq j$, they are orthogonal since they correspond to distinct characteristic roots (Lemma 9–5). Thus $\xi_1, \ldots, \xi_n$ form an orthonormal basis for $V_n(\Re)$ consisting of characteristic vectors of A, and this proves the theorem by virtue of the criterion in Theorem 9–7.

If we order the r_i, all real numbers by Theorem 9–9, so that

$$r_1 \leqq r_2 \leqq \cdots \leqq r_n,$$

there would then be a unique diagonal matrix similar to each given real, symmetric matrix. In other words, we have found a canonical set for real, symmetric matrices under orthogonal similarity.

The reader would do well to learn the proof of Theorem 9–11. The same general direction will be followed later in this chapter when Hermitian matrices are studied and, with some modifications, when normal matrices are studied.

THEOREM 9–12. *Two real, symmetric matrices are orthogonally similar if and only if they have the same characteristic roots, hence, if and only if they are similar.*

The proof is left to the reader.

EXERCISES

1. Prove Theorem 9–12.

2. If A_1 and A_2 denote the first and last matrices, respectively, in Exercise **2**, Section 9–7, find orthogonal matrices P_i such that $P_i^{-1} A_i P_i$ is diagonal, $i = 1, 2$. Then find an orthogonal matrix Q such that $Q^{-1} A_1 Q = A_2$.

3. Prove that the real, symmetric matrices are the only real matrices which are orthogonally similar to diagonal matrices.

9-9 Pairs of quadratic forms. One of the interesting and useful applications of orthogonal similarity is concerned with the simultaneous reduction of two quadratic forms in $x_1, \ldots, x_n$, that is, a simplification of both forms by the same substitution $X = PY$. Stated in matrix form the result is

THEOREM 9-13. *Let A and B be $n \times n$ real, symmetric matrices, A being positive definite. Then there is a nonsingular, real matrix R such that*

(23) $R'AR = I, \quad R'BR = \text{diag}(r_1, \ldots, r_n).$

For any choice of R the quantities $r_1, \ldots, r_n$ are necessarily the roots of the polynomial equation, $|xA - B| = 0$.

There is (Theorem 5-8) a nonsingular, real matrix Q such that $Q'AQ = I$. Then $C = Q'BQ$ is a real, symmetric matrix, whence (Theorem 9-11) there is an orthogonal matrix P such that

$$P'CP = \text{diag}(r_1, \ldots, r_n) = D, \quad P' = P^{-1}.$$

Then $R = QP$ fulfills (23). Also

$$R'(xA - B)R = xI - D.$$

But $|R| = r \neq 0$, where r is a real number, and

$$|xI - D| = |R|^2 \cdot |xA - B| = r^2 \cdot |xA - B|,$$

whence the roots $r_1, \ldots, r_n$ of $|xI - D| = 0$ are also the roots of $|xA - B| = 0$.

EXERCISES

1. Find the roots of the equation $|xA - B| = 0$, where A and B are the real matrices

$$A = \begin{bmatrix} 1 & 1 \\ 1 & 4 \end{bmatrix}, \quad B = \begin{bmatrix} 0 & 3 \\ 3 & 0 \end{bmatrix}.$$

2. Using the results of Exercise 1, show that the real quadratic forms

$$f = x_1^2 + 2x_1x_2 + 4x_2^2, \quad g = 6x_1x_2$$

may be simultaneously reduced by a nonsingular linear substitution to $y_1^2 + y_2^2$ and $y_1^2 - 3y_2^2$.

3. Prove Theorem 9-11 as a corollary of Theorem 9-13.

4. If A and B are $n \times n$ matrices and A is nonsingular, show that $A^{-1}B$ and BA^{-1} have the same characteristic roots. How are these roots related to the roots of $|xA - B|$?

9–10 Inner products over ℂ. The theory in Chapter 5 of congruent symmetric matrices over the real number field had its counterpart over the complex number field in the theory of Hermitely congruent Hermitian matrices. Recall that matrices B and A over ℂ are Hermitely congruent if there is a nonsingular complex matrix P such that

$$(24) \qquad B = \bar{P}'AP.$$

The orthogonal similarity of real, symmetric matrices also has its counterpart over ℂ: the *unitary similarity* of Hermitian matrices. Any two matrices B and A over ℂ are called *unitarily similar* if (24) holds with a matrix P such that

$$(25) \qquad \bar{P}' = P^{-1}.$$

Note that a real matrix obeying (25) is orthogonal and that (24) and (25) give an orthogonal similarity if all the matrices concerned are real.

In order to study unitary similarity, we begin by generalizing the concepts which were central in the study of orthogonal similarity. We shall extend the concepts of inner product, length, and orthogonality so that they apply to vectors in $V_n(ℂ)$. As usual, let $\bar{c}$ denote the complex conjugate $a - bi$ of a complex number $c = a + bi$.

DEFINITION 5. If $\xi = \mathrm{col}\ (a_1, \ldots, a_n)$ and $\eta = \mathrm{col}\ (b_1, \ldots, b_n)$ belong to $V_n(ℂ)$, the *inner product* of ξ with η is defined to be

$$\bar{\xi}'\eta = \bar{a}_1 b_1 + \cdots + \bar{a}_n b_n.$$

Warning: Unlike inner products of real vectors, the inner product $\bar{\eta}'\xi$ of η with ξ is not the same as that of ξ with η. The two are complex conjugates:

$$\bar{\eta}'\xi = \bar{b}_1 a_1 + \cdots + \bar{b}_n a_n = \overline{\bar{\xi}'\eta}.$$

If $c = a + bi$, the product of c with its conjugate $\bar{c} = a - bi$ is

$$\bar{c}c = a^2 + b^2,$$

hence is a positive real number if $c \neq 0$. It follows that the inner product

$$\bar{\xi}'\xi = \bar{a}_1 a_1 + \cdots + \bar{a}_n a_n$$

of ξ with itself is positive if $\xi \neq 0$, and is zero if $\xi = \mathbf{0}$.

DEFINITION 6. If ξ is in $V_n(ℂ)$, the *length* of ξ is the nonnegative square root

$$(\bar{\xi}'\xi)^{\frac{1}{2}}.$$

A vector is called *normal* if its length is unity.

If $\bar{\xi}'\eta = 0$, we find on taking the conjugate transpose that $\bar{\eta}'\xi = 0$. This clears the path for the concept of orthogonality in $V_n(\mathbb{C})$.

DEFINITION 7.   Two vectors in $V_n(\mathbb{C})$ are called *orthogonal* if their inner product (in either order) is zero.

Each $n \times n$ matrix A over $\mathbb{C}$ determines a linear transformation

$$(26) \qquad\qquad \eta = A\xi$$

of $V_n(\mathbb{C})$ in which each vector ξ is "transformed" into the new vector (26).

DEFINITION 8.   An $n \times n$ matrix A over $\mathbb{C}$ is called *unitary* if the linear transformation (26) determined by A preserves lengths:

$$(\bar{\eta}'\eta)^{\frac{1}{2}} = (\bar{\xi}'\xi)^{\frac{1}{2}}$$

for every vector ξ of $V_n(\mathbb{C})$. The linear transformation (26) is also called unitary in this case.

Since

$$\bar{\eta}'\eta = (\overline{A\xi})'(A\xi) = \bar{\xi}'\bar{A}'A\xi,$$

it follows as in the proof of Theorem 9–6 that A is unitary if and only if $\bar{A}'A = I$. This equation and its row-by-column interpretation give

THEOREM 9–14.   *A square matrix A over $\mathbb{C}$ is unitary if and only if $\bar{A}' = A^{-1}$. This is true, also, if and only if the columns A_i of A are mutually orthogonal, normal vectors:*

$$\bar{A}'_i A_i = 1, \quad \bar{A}'_i A_j = 0. \qquad\qquad (i \neq j)$$

Just as in the real case, orthogonality implies linear independence.

LEMMA 9–7.   *If $\xi_1, \ldots, \xi_t$ are mutually orthogonal, nonzero vectors of $V_n(\mathbb{C})$, they are linearly independent.*

The proof proceeds exactly as for Lemma 9–4, conjugate transposes now being employed where transposes were used in the earlier case.

DEFINITION 9.   A *normal unitary basis* of $V_n(\mathbb{C})$ is a basis consisting of vectors which are normal and mutually orthogonal.

Thus a matrix is unitary if and only if its columns form a normal unitary basis for $V_n(\mathbb{C})$. One such basis for $V_n(\mathbb{C})$ is the basis of unit vectors. Every subspace also has such a basis, as we shall see.

THEOREM 9–15. *Every nonzero vector space V over $\mathfrak{C}$ has a normal unitary basis. Every set of mutually orthogonal, normal vectors of V may be extended to a normal unitary basis of V.*

The proof is completely parallel to that of Theorem 9–8 and its corollary. Details are left as an exercise.

THEOREM 9–16. *An $n \times n$ matrix A over $\mathfrak{C}$ is unitarily similar to a diagonal matrix if and only if there is a normal unitary basis of $V_n(\mathfrak{C})$ consisting of characteristic vectors of A.*

The proof is almost word for word the same as that for Theorem 9–7. The matrix P such that $\bar{P}'AP = \mathrm{diag}\,(r_1, \ldots, r_n)$ has for its columns the characteristic vectors described in Theorem 9–16. This matrix P is unitary.

EXERCISES

★1. Show that unitary similarity is an RST relation.

★2. If A is unitary, prove that every matrix unitarily similar to A is unitary.

3. If A is Hermitian, show that every matrix unitarily similar to A is Hermitian.

★4. If A is unitary and $a = |A|$, prove that $a\bar{a} = 1$ and A^{-1}, A', and $\bar{A}$ are unitary.

★5. If a vector β is orthogonal to each of the vectors $\alpha_1, \ldots, \alpha_h$, show that β is orthogonal to every linear combination of $\alpha_1, \ldots, \alpha_h$.

6. Prove Lemma 9–7.

7. Prove Theorem 9–15.

8. Prove Theorem 9–16.

9. Prove that a product of unitary matrices is always unitary.

10. If the columns of a unitary matrix are permuted in any way, show that the resulting matrix is unitary.

11. Show that the diagonal elements of the matrix in Theorem 9–16 may be made to appear on the diagonal in any desired order.

12. Let $B = P^{-1}AP$, where P is unitary. Show that all unitary matrices Q obeying $Q^{-1}AQ = B$ are given by $Q = RP$, where R varies over all unitary matrices commuting with A.

9–11 Unitary similarity of Hermitian matrices. We are now prepared to follow a path entirely parallel to that which led through Sections 9–7 and 9–8 to Theorem 9–11. The present path is associated with the scalar field $\mathfrak{C}$ instead of $\mathfrak{R}$ and leads to Theorem 9–17. Details of proof in most cases are left as exercises, since the proofs already given for the real case require only slight modification.

LEMMA 9–8. *Every Hermitian matrix A is similar to a diagonal matrix (whose diagonal elements are necessarily the characteristic roots of A).*

Since (Theorem 9–9) the characteristic roots of A are real, the minimum polynomial of A has real coefficients. With this start we may easily paraphrase the proof of Lemma 9–6 so that it proves Lemma 9–8.

THEOREM 9–17. *Every Hermitian matrix A is unitarily similar to a diagonal matrix (whose diagonal elements are necessarily the characteristic roots of A).*

The proof, parallel to that of Theorem 9–11, is left to the reader. Unitary similarity is an RST relation, and the famous result just stated amounts to the determination of a canonical set of Hermitian matrices relative to this relation. Theorem 9–12 and its proof may likewise be paralleled:

THEOREM 9–18. *Two Hermitian matrices are unitarily similar if and only if they have the same characteristic roots; also, if and only if they are similar.*

Minor changes in the proof of Theorem 9–13 lead to

THEOREM 9–19. *Let A and B be n × n Hermitian matrices, A being positive definite. Then there is a nonsingular matrix R over $\mathbb{C}$ such that*

$$R'AR = I, \quad R'BR = \text{diag}(r_1, \ldots, r_n).$$

For any choice of R the quantities $r_1, \ldots, r_n$ are necessarily the roots of the polynomial equation $|xA - B| = 0$.

EXERCISES

1. Interpret Theorem 9–17 for Hermitian forms.
2. Write detailed proofs of (a) Lemma 9–8, (b) Theorem 9–18, (c) Theorem 9–19.

9–12 Normal matrices. Those matrices over a general field $\mathfrak{F}$ which are similar to diagonal matrices have been characterized fully in Section 9–2. Those real matrices which are orthogonally similar to diagonal matrices are simply the symmetric matrices (why?). Contemplating the complex number field as the field of scalars, we now inquire which matrices are unitarily similar to diagonal matrices. This class of matrices includes all Hermitian matrices (Theorem 9–17), and a natural suggestion is that the class includes nothing but Hermitian matrices. Actually, as we shall see (Theorems 9–21, 9–23), there are many other matrices which are unitarily similar to diagonal matrices. We shall find a simple characterization of all of them.

DEFINITION 10. A complex, square matrix A is called *normal* if it commutes with its conjugate transpose: $A\bar{A}' = \bar{A}'A$.

Suppose that

$$\bar{P}'AP = D = \text{diag}\ (r_1, \ldots, r_n), \quad P^{-1} = \bar{P}'.$$

Then, taking conjugate transposes, we find that

$$\bar{P}'\bar{A}'P = \bar{D}' = \bar{D} = \text{diag}\ (\bar{r}_1, \ldots, \bar{r}_n).$$

Since $D\bar{D} = \bar{D}D$, it follows that

$$P^{-1}AP \cdot P^{-1}\bar{A}'P = P^{-1}\bar{A}'P \cdot P^{-1}AP,$$
$$P^{-1}A\bar{A}'P = P^{-1}\bar{A}'AP,$$

so that
$$A\bar{A}' = \bar{A}'A.$$

This proves

LEMMA 9–9. *If A is unitarily similar to a diagonal matrix, it is normal.*

In proving the converse of this lemma we shall follow a path like that in the orthogonal reduction of real, symmetric matrices and in its Hermitian analogue. Observe first that if A is normal, so is $A - rI$ for any complex number r.

LEMMA 9–10. *If A is normal and ξ is a characteristic vector of A corresponding to r, then it also is a characteristic vector of $\bar{A}'$ corresponding to $\bar{r}$.*

Let $B = A - rI$ so that $B\bar{B}' = \bar{B}'B$. Then $B\xi = 0$ and

$$0 = (\overline{B\xi})'(B\xi) = \bar{\xi}'\bar{B}'B\xi = (\bar{\xi}'B)(\bar{B}'\xi)$$
$$= \bar{\eta}'\eta, \quad \eta = \bar{B}'\xi.$$

Since the inner product of a vector with itself is zero only if the vector is zero, it follows that $\eta = 0$,

$$\bar{B}'\xi = 0 = (\bar{A}' - \bar{r}I)\xi, \quad \bar{A}'\xi = \bar{r}\xi.$$

This result makes it possible to prove a generalization of **Lemma 9–5:**

THEOREM 9–20. *If A is normal, characteristic vectors of A corresponding to distinct characteristic roots are orthogonal.*

We have $r_1 \neq r_2$ and

$$A\xi_i = r_i\xi_i. \qquad (i = 1, 2)$$

Then $\bar{A}'\xi_1 = \bar{r}_1\xi_1$ (Lemma 9–10), so that

$$\begin{aligned}
\bar{\xi}'_1(A\xi_2) &= r_2\bar{\xi}'_1\xi_2 \\
&= (\bar{\xi}'_1 A)\xi_2 = (\overline{\bar{A}'\xi_1})'\xi_2 \\
&= (\overline{\bar{r}_1\xi_1})'\xi_2 = r_1\bar{\xi}'_1\xi_2,
\end{aligned}$$

$$(r_1 - r_2)(\bar{\xi}'_1\xi_2) = 0, \quad \bar{\xi}'_1\xi_2 = 0,$$

since $r_1 - r_2$ is not zero.

LEMMA 9–11. *If r is a characteristic root of multiplicity k of a normal matrix A, the null space of $rI - A$ has dimension k.*

This result is a crucial step towards the canonical set. It will follow as soon as we show that A is similar to a diagonal matrix, and this in turn will follow from the fact, now to be proved, that the minimum polynomial $m(x)$ of A has no repeated factor. Suppose, to the contrary, that

$$m(x) = (x - r)^2 h(x) = (x - r)g(x)$$

so

$$g(x) = (x - r)h(x), \quad g(A) = (A - rI)h(A),$$
$$m(A) = 0 = (A - rI)g(A), \quad g(A) \neq 0.$$

Since $g(A)$ is a nonzero matrix there is a vector η such that

$$(27) \qquad\qquad g(A)\eta = \xi \neq 0.$$

This vector ξ is actually a characteristic vector of A corresponding to r, as the following computation shows:

$$(A - rI)\xi = (A - rI)g(A)\eta = m(A)\eta = 0.$$

By Lemma 9–10,

$$(\bar{A}' - \bar{r}I)\xi = 0.$$

Conjugate transposes in this equation give

$$\bar{\xi}'(A - rI) = 0,$$
$$\bar{\xi}'\xi = \bar{\xi}'g(A)\eta = \bar{\xi}'(A - rI)h(A)\eta$$
$$= 0 \cdot h(A) = 0.$$

Thus $\bar{\xi}'\xi = 0$ so $\xi = 0$. This conflict with (27) establishes the lemma.

With this lemma we are able to prove the main result on normal matrices.

THEOREM 9–21. *A complex matrix is unitarily similar to a diagonal matrix if and only if it is normal.*

As in the case of corresponding results for real, symmetric matrices and for Hermitian matrices, we consider the distinct characteristic roots $s_1, \ldots, s_t$ of the normal matrix A, and the null space N_i of $s_iI - A$ $(i = 1, \ldots, t)$. By the lemma just proved, the dimension m_i of N_i is the multiplicity of s_i, so

$$n = m_1 + \cdots + m_t.$$

Let $\xi_1, \ldots, \xi_n$ be the vectors obtained by putting together any chosen normal unitary bases for $N_1, N_2, \ldots, N_t$. Then ξ_i and ξ_j are orthogonal if they belong to the same N_k, and also if they belong to different null spaces N_h and N_k, by Theorem 9–20. Hence the normal vectors $\xi_1, \ldots, \xi_n$ are mutually orthogonal, thus linearly independent (Lemma 9–7), and so form a normal unitary basis for $V_n(\mathcal{C})$.

Since the normal unitary basis obtained in this way consists of characteristic vectors of A, Theorem 9–16 implies that every normal matrix is unitarily similar to a diagonal matrix. No other matrices have this property, by Lemma 9–9, and the proof is complete.

COROLLARY 9–21. *If A is normal, its principal idempotents $E_1, \ldots, E_t$ (see Theorem 9–5) are Hermitian. Conversely, if A is a diagonable complex matrix whose principal idempotents are all Hermitian, A is normal.*

On examining the proof of Theorem 9–5 and using the fact that P is now unitary, we find that $E_i = PD_iP^{-1}$ is Hermitian. Conversely, if the E_i are all Hermitian,

$$A = s_1E_1 + \cdots + s_tE_t,$$
$$\bar{A}' = \bar{s}_1E_1 + \cdots + \bar{s}_tE_t,$$

so $A\bar{A}' = s_1\bar{s}_1E_1 + \cdots + s_t\bar{s}_tE_t = \bar{A}'A$ and A is normal.

The diagonal elements obtained in Theorem 9–21 must be the characteristic roots of the given normal matrix. This gives the following parallel of Theorems 9–12 and 9–18.

THEOREM 9–22. *Two normal matrices are unitarily similar if and only if they have the same characteristic roots.*

The class of normal matrices includes a wide variety.

THEOREM 9-23. *All matrices of the following types are normal: Hermitian, skew-Hermitian, real symmetric, real skew, unitary, orthogonal, diagonal, and all matrices unitarily similar to normal matrices.*

The routine verification is left to the reader. A few warnings concerning the above theorems are in order. First, there are matrices which are normal but do not belong to any of the types in Theorem 9–23 (except the all-inclusive last type). For example,

$$A = \begin{bmatrix} 0 & 1+i \\ 1+i & 0 \end{bmatrix}.$$

Second, if Theorem 9–21 is applied to a normal, real matrix, the unitary similarity does not necessarily become an orthogonal similarity. In fact a real skew matrix $\neq 0$ is by the last two results unitarily similar to a diagonal matrix. But it is not orthogonally similar to a diagonal matrix, since this property is enjoyed by a real matrix A only if A is symmetric.

EXERCISES

1. Prove that a 2×2 complex matrix

$$\begin{bmatrix} 0 & a \\ b & 0 \end{bmatrix}$$

with zero diagonal is normal if and only if $a\bar{a} = b\bar{b}$.

2. Prove that every 2×2 complex matrix of the form

$$\begin{bmatrix} a & c \\ c & a \end{bmatrix}$$

is normal.

3. Show how the proof of Theorem 9–21 employs Lemma 9–11.

★4. For every integer $k > 0$ and every normal matrix A show that there is a normal matrix B such that B has the same rank as A and $B^k = A$. Hint: Use Theorem 9–21.

★5. Let A be a real, symmetric matrix which is positive semidefinite. Show that for every integer $k > 0$ there is a real, positive semidefinite matrix B of the same rank as A such that $B^k = A$.

6. Prove Theorem 9–23.

7. Prove that every normal matrix is similar to its transpose.

8. Show that two normal matrices are unitarily similar if and only if they are similar.

9. If A is normal, show that $A\xi$ and $\bar{A}'\xi$ always have the same length.

10. If A is normal and $g(x)$ is any polynomial over $\mathcal{C}$, prove that $g(A)$ is normal.

11. If A is normal and nonsingular, prove that A^{-1} is normal.

9–13 Characteristic roots of certain normal matrices. Various types of normal matrices are easily distinguished by their characteristic roots, as indicated in Theorem 9–25.

THEOREM 9–24. *If A is unitary, its characteristic roots have absolute value one. If A is skew-Hermitian, its nonzero characteristic roots are pure imaginaries.*

In each case let ξ be a nonzero characteristic vector corresponding to r:

$$A\xi = r\xi.$$

If A is unitary, $\bar{A}'A = I$, so that

$$\xi = \bar{A}'A\xi = r\bar{A}'\xi = r\bar{r}\xi$$

by Lemma 9–10. Then

$$(1 - r\bar{r})\xi = 0, \quad \xi \neq 0, \quad r\bar{r} = 1.$$

Thus the absolute value $(r\bar{r})^{\frac{1}{2}}$ of r is 1.

If A is skew-Hermitian, $\bar{A}' = -A$,

$$\begin{aligned}
\bar{\xi}'A\xi &= r\bar{\xi}'\xi \\
&= -\bar{\xi}'\bar{A}'\xi = -(\overline{A\xi})'\xi \\
&= -(\overline{r\xi})'\xi = -\bar{r}\bar{\xi}'\xi, \\
r(\bar{\xi}'\xi) &= -\bar{r}(\bar{\xi}'\xi), \quad \bar{\xi}'\xi \neq 0, \quad \bar{r} = -r.
\end{aligned}$$

The last equation shows that r is a pure imaginary complex number if it is nonzero.

THEOREM 9–25. *Let A be a normal matrix with characteristic roots $r_1, \ldots, r_n$. Then each of the following properties (j) of A is equivalent to the property (j') of the characteristic roots $(j = a, b, c, d, e, f)$:*

(a) *A is Hermitian;* (a') *every r_i is real;*

(b) *A is positive definite;* (b') *every r_i is real and > 0;*

(c) *A is positive semidefinite;* (c') *every r_i is real and $\geqq 0$;*

(d) *A is nonsingular;* (d') *every $r_i \neq 0$;*

(e) *A is unitary;* (e') *every r_i has $|r_i| = 1$;*

(f) *A is skew-Hermitian;* (f') *every nonzero r_i is a pure imaginary.*

In every case

$$\begin{aligned}
\bar{P}'AP = D &= \text{diag}\,(r_1, \ldots, r_n) \\
&= P^{-1}AP, \quad P^{-1} = \bar{P}'.
\end{aligned}$$

Then (d) implies (d') by preservation of rank. That (e) and (f) imply (e') and (f'), respectively, is asserted in the previous theorem.

Conversely, if (d') holds, D is nonsingular, whence so is $A = PDP^{-1}$. If all r_i have absolute value 1,

$$r_i \bar{r}_i = 1, \quad r_i^{-1} = \bar{r}_i,$$

so that $D^{-1} = \bar{D} = \bar{D}'$, D is unitary, and so is the unitarily similar matrix A. If (f') holds, D is skew-Hermitian, and so is every matrix Hermitely congruent to it, including A. The first three of the six results above are left as exercises.

A property of positive definite Hermitian matrices was given in the second part of Theorem 5–18. In Theorem 9–26 we shall prove the converse of this property.

LEMMA 9–12. *If A is any square matrix, the polynomial $g(x)$ $= |A - xI|$ has the form*

(28) $$g(x) = (-x)^n + c_{n-1}(-x)^{n-1} + \cdots + c_1(-x) + c_0,$$

where c_k is the sum of all principal subdeterminants of $|A|$ having $n - k$ rows $(k = 0, 1, \ldots, n - 1)$.

Each term involving x^k in the expansion of $|A - xI|$ is found by selecting k diagonal elements $a_{ii} - x$, say $i = j_1, \ldots, j_k$, and taking from the product,

$$\prod_{i=j_1}^{j_k} (a_{ii} - x),$$

only the term in x^k, namely

$$(-x)^k = (-1)^k x^k.$$

The remaining $n - k$ factors must be chosen in arbitrary fashion from rows and columns not yet used, that is, from the principal subdeterminant $|S|$ obtained by crossing out the rows and columns numbered $j_1, \ldots, j_k$. Since we want a term in x^k, we must choose the constant term from $|S|$. This is equivalent to replacing x by 0 in $|S|$, and yields the principal subdeterminant of A resulting from deletion of the rows and columns numbered $j_1, \ldots, j_k$. All possible terms in $(-x)^k$ are found by choosing $j_1, \ldots, j_k$ in all possible ways. This gives the lemma.

THEOREM 9–26. *Let A be Hermitian. Then A is positive definite if and only if $|A|$ and all principal subdeterminants of A are positive.*

The necessity of this condition is asserted in Theorem 5–18. Conversely, if $|A|$ and all principal subdeterminants are positive, all the

c_k in (28) are positive, so that $g(-x)$ is a polynomial in x with no variation in the signs of its coefficients. By Descartes' rule* of signs $g(x)$ has no negative roots. It has no zero roots, since $c_0 = |A| > 0$. Therefore every real root is positive. But

$$f(x) = |xI - A| = (-1)^n|A - xI| = (-1)^n g(x),$$

and the roots of $g(x)$ are the characteristic roots of A, which are known to be real. Since we have just proved that they are not zero or negative, all the roots are positive, and Theorem 9–25 implies that A is positive definite.

EXERCISES

1. Prove the first three parts of Theorem 9–25.

2. Prove that a normal matrix A is idempotent if and only if each of its characteristic roots is zero or unity. Find a nonidempotent matrix whose characteristic roots are 1, 1, 0.

3. Determine which of the following Hermitian matrices are positive definite:

$$\begin{bmatrix} 1 & 2 \\ 2 & 3 \end{bmatrix}, \quad \begin{bmatrix} 1 & i & 0 \\ -i & 2 & 1+i \\ 0 & 1-i & 3 \end{bmatrix}.$$

4. Prove that $cI + A$ is nonsingular if

(a) A is skew-Hermitian and c is any complex number which is not a pure imaginary;

(b) A is Hermitian and c is any imaginary complex number;

(c) A is positive semidefinite and c is either imaginary or a real positive number;

(d) A is unitary and c is any complex number whose absolute value is different from unity.

5. In Exercise 4, Section 9–12, show that B is necessarily unitary if A is unitary, and that B can be chosen to be positive semidefinite if A has this property.

9–14 Normal, real matrices. If A is normal, there is a unitary matrix U, not unique, such that

$$(29) \qquad U^{-1}AU = \text{diag}(r_1, \ldots, r_n),$$

where the r_i are the characteristic roots of A. This theorem provides a canonical set of normal matrices under unitary similarity. When A

---

* See any text in College Algebra or Theory of Equations, in particular Reference 16.

is real as well as normal, it is natural to consider what happens if the unitary matrix U is restricted to being real, hence orthogonal. We shall find a canonical set of normal, real matrices relative to orthogonal similarity. We cannot expect to find diagonal matrices as in (29), since under present assumptions the left side of (29) is real, whereas some of the characteristic roots r_i may be imaginary. If A is real and skew, for example, it is normal, but all of its nonzero characteristic roots are imaginary.

Since A is real, its characteristic polynomial $f(x)$ has real coefficients. Hence if $f(x)$ has any imaginary roots they fall into pairs of conjugate imaginaries

$$(30) \qquad\qquad a_j \pm b_j i. \qquad\qquad (j = 1, \ldots, m)$$

We shall show that A is orthogonally similar to a modification of the diagonal matrix (29), in which each pair of conjugate imaginaries (30) is replaced by the diagonal block

$$(31) \qquad\qquad C_j = \begin{bmatrix} a_j & -b_j \\ b_j & a_j \end{bmatrix}.$$

THEOREM 9-27. *Let A be a normal, real matrix whose nonreal characteristic roots are the imaginary numbers* (30), *and whose real characteristic roots are $r_1, \ldots, r_q$. Then A is orthogonally similar to*

$$B = \text{diag } (R, C_1, \ldots, C_m),$$

where the C_j are defined by (31), (30), *and $R = \text{diag } (r_1, \ldots, r_q).$*[*]

If the proof of Theorem 9-17 is well in mind, it will be easy to follow the modifications which will be made to prove the present theorem. First we prove some lemmas in which A always means the matrix of Theorem 9-27.

LEMMA 9-13. *If M is an $n \times n$ real matrix, any basis for the null space over $\mathfrak{R}$ of M is also a basis for the null space over $\mathfrak{C}$ of M. Hence the latter space N has a basis of normal, mutually orthogonal, real vectors.*

The rank g of M is unchanged by enlarging the scalar field from $\mathfrak{R}$ to $\mathfrak{C}$ (Corollary 3-11). Then the null space of M over both $\mathfrak{R}$ and $\mathfrak{C}$ has dimension $h = n - g$. If $\alpha_1, \ldots, \alpha_h$ form an orthonormal basis for the null space of M over $\mathfrak{R}$, these vectors belong to the null space

---

[*] If there are no real characteristic roots, the block R is absent from B. and if all roots are real, R is the only block present.

of M over $\mathcal{C}$ and are linearly independent. (Why? Find a relevant exercise in Chapter 3.) These facts complete the proof.

LEMMA 9–14. *Let $a \pm bi$ $(b \neq 0)$ be a pair r, $\bar{r}$, of conjugate imaginary characteristic roots of A. Let N and $\bar{N}$ be the null spaces (over $\mathcal{C}$) of $rI - A$ and $\bar{r}I - A$, respectively. Then if $\alpha_1, \ldots, \alpha_h$ form a basis for N, $\bar{\alpha}_1, \ldots, \bar{\alpha}_h$ form a basis for $\bar{N}$.*

The conjugate $\bar{\alpha}$ of α is the vector whose coordinates are the conjugates of those of α. Since A is real, $A\alpha_i = r\alpha_i$ implies

$$\overline{A\alpha_i} = \overline{r\alpha_i}, \quad A\bar{\alpha}_i = \bar{r}\bar{\alpha}_i,$$

so that α_i belongs to $\bar{N}$. The independence of $\alpha_1, \ldots, \alpha_h$ implies that of $\bar{\alpha}_1, \ldots, \bar{\alpha}_h$, whence the dimension $\bar{h}$ of $\bar{N}$ obeys: $\bar{h} \geq h$. Since the roles of N and $\bar{N}$ are interchangeable, $h \geq \bar{h}$, $h = \bar{h}$, so that $\bar{\alpha}_1, \ldots, \bar{\alpha}_h$ form a basis of $\bar{N}$.

LEMMA 9–15. *Let ξ and $\bar{\xi}$ be conjugate characteristic vectors of A corresponding to $a + bi$ and $a - bi$, respectively $(b \neq 0)$. Let*

$$\gamma = \xi + \bar{\xi}, \quad \delta = i(\xi - \bar{\xi}).$$

Then

    (a) γ *and* δ *are real, nonzero vectors;*
    (b) $A\gamma = a\gamma + b\delta$, $A\delta = -b\gamma + a\delta$;
    (c) $A'\gamma = a\gamma - b\delta$, $A'\delta = b\gamma + a\delta$;
    (d) γ *is orthogonal to* δ;
    (e) γ *and* δ *have the same length.*

The reality of γ and δ is clear. If γ were zero, ξ would have to be pure imaginary (i.e., $\xi = i\eta$ where η is real) and if δ were zero, ξ would have to be real. But

(32) $A\xi = (a + bi)\xi = a\xi + bi\xi, \quad b \neq 0.$

If ξ is real, the left side of (32) is real and the right side is not; if ξ is pure imaginary, $\xi = i\eta$, the left side is pure imaginary, and the right side, $ai\eta - b\eta$, is not. Thus γ and δ are not zero, and (a) is proved.

Next we consider (c), and use the fact that $A' = \bar{A}'$. Let $a + bi = r$. By Lemma 9–10

$$A'\xi = \bar{r}\xi, \quad \overline{A'\xi} = A'\bar{\xi} = r\bar{\xi},$$
$$A'\gamma = \bar{r}\xi + r\bar{\xi} = (a - bi)\xi + (a + bi)\bar{\xi}$$
$$= a(\xi + \bar{\xi}) - bi(\xi - \bar{\xi})$$
$$= a\gamma - b\delta.$$

The second part of (c) and both parts of (b) are derived similarly.
For (d) we compute

$$\begin{aligned}
\gamma'A\gamma &= \gamma'(a\gamma + b\delta) = a(\gamma'\gamma) + b(\gamma'\delta) \\
&= (A'\gamma)'\gamma = (a\gamma - b\delta)'\gamma \\
&= a(\gamma'\gamma) - b(\delta'\gamma).
\end{aligned}$$

Therefore, $b(\gamma'\delta) = -b(\delta'\gamma)$. Since $b \neq 0$ and $\delta'\gamma = (\delta'\gamma)' = \gamma'\delta$ we obtain $\gamma'\delta = -\gamma'\delta$, whence $\gamma'\delta = 0$. This proves (d). To prove (e) we compute

$$\begin{aligned}
\gamma'A\delta &= \gamma'(-b\gamma + a\delta) = -b(\gamma'\gamma) \\
&= (A'\gamma)'\delta = (a\gamma - b\delta)'\delta \\
&= -b(\delta'\delta), \qquad b \neq 0.
\end{aligned}$$

Therefore $\gamma'\gamma = \delta'\delta$. This completes the proof of the lemma.

Now we are ready for the proof of Theorem 9-27. To find an orthogonal matrix P such that $P^{-1}AP = B$ or, equivalently,

$$AP = PB,$$

it is sufficient (and necessary) to find a set

(33) $$\eta_1, \ldots, \eta_q, \gamma_1, \delta_1, \ldots, \gamma_m, \delta_m$$

of mutually orthogonal, normal vectors of $V_n(\mathfrak{R})$ such that

(34) $$A\eta_k = r_k\eta_k, \qquad\qquad (k = 1, \ldots, q)$$

(35) $$\begin{cases} A\gamma_j = a_j\gamma_j + b_j\delta_j \\ A\delta_j = -b_j\gamma_j + a_j\delta_j. \end{cases} \qquad (j = 1, \ldots, m)$$

Since the last pair of equations may be written as

$$A(\gamma_j, \delta_j) = (\gamma_j, \delta_j)\begin{bmatrix} a_j & -b_j \\ b_j & a_j \end{bmatrix},$$

it is clear that the matrix whose columns are the vectors (33) will do the job required of P.

To find the vectors (33) consider, as in the proof of Theorem 9-17, the null spaces $N(s)$ over $\mathcal{C}$ of the matrices $sI - A$, s varying over the distinct characteristic roots of A. If s is real, $N(s)$ has, by Lemma 9-13, a basis of real vectors which may be chosen to be normal and mutually orthogonal. The number of vectors in this basis is (Lemma 9-11) the number of r_i which are equal to s. Making such a choice of basis for each $N(s)$ for which s is real, we obtain the desired vectors $\eta_1, \ldots, \eta_q$, mutual orthogonality being assured by Theorem 9-20.

If $s = a + bi$ is an imaginary root, and if $\alpha_1, \ldots, \alpha_h$ form a normal unitary basis for $N(s)$, then $\bar{s} = a - bi$ is also a root and $N(\bar{s})$ has

$\bar{\alpha}_1, \ldots, \bar{\alpha}_h$ as a basis (Lemma 9–14) which also is normal and unitary. (Why?)   Let ξ denote any one of the α_j, so that $\bar{\xi} = \bar{\alpha}_j$.   Then ξ and $\bar{\xi}$ may be used in Lemma 9–15 to define vectors γ and δ which obey equations (b).   Since by (e) γ and δ have the same length l, division by l converts equations (b) into equations (35) in which

$$\gamma_j = \gamma/l, \quad \delta_j = \delta/l$$

are normal.   The orthogonality of these two vectors is implied by (d).

For a different pair of conjugate imaginary roots there will, likewise, be basis vectors $\alpha_{h+1}, \ldots, \alpha_k$ and $\bar{\alpha}_{h+1}, \ldots, \bar{\alpha}_k$.   Continuing for all the distinct pairs $s, \bar{s}$ we obtain $2m$ vectors which are mutually orthogonal:

$$(36) \qquad \alpha_1, \bar{\alpha}_1, \ldots, \alpha_m, \bar{\alpha}_m.$$

Replacing each pair α_j, $\bar{\alpha}_j$ by γ_j, δ_j as above gives the set of vectors

$$\gamma_1, \delta_1, \ldots, \gamma_m, \delta_m,$$

which are normal and satisfy (35).

It remains to verify the orthogonality of the set (33), this having been done for the first q vectors η_k and for each pair γ_j, δ_j.   Since η_k and either of α_j, $\bar{\alpha}_j$ are characteristic vectors of A corresponding to two distinct roots, η_k is orthogonal to every vector in (36), hence also to their linear combinations γ_j, δ_j.   Now let μ_j denote either γ_j or δ_j and μ_k denote γ_k or δ_k $(j \neq k)$.   Since α_k is orthogonal to both α_j and $\bar{\alpha}_j$, it is orthogonal to μ_j.   Likewise $\bar{\alpha}_k$ is orthogonal to μ_j, and since μ_j is orthogonal to both α_k and $\bar{\alpha}_k$, μ_j is orthogonal to μ_k.   This completes the proof.   The proof of the next statement is left as an exercise.

THEOREM 9–28.   *Two real, normal matrices are orthogonally similar if and only if they have the same characteristic roots; also, if and only if they are similar.*

EXERCISES

1. Prove Theorem 9–28.

2. Prove the linear independence of the vectors $\bar{\alpha}_1, \ldots, \bar{\alpha}_h$ in Lemma 9–14.

3. Using Theorem 9–25 show in detail how Theorem 9–27 specializes in case A is a real, skew matrix or an orthogonal matrix.

4. Give an example of a real, normal matrix which is neither orthogonal, skew, nor symmetric.

5. Lemma 9–14 implies that the imaginary roots of a normal, real matrix occur in conjugate pairs $a \pm bi$ such that $a - bi$ has the same multiplicity as $a + bi$.   Is this true for normal matrices which are not real?

6. Prove that Theorem 9–27 remains valid if some or all of the matrices C_j are replaced by their transposes.

7. Find a proof based only on the ideas of Chapter 8 that the matrices A and B of Theorem 9–27 are similar.

9–15 An interesting factorization. It is a remarkable fact that an arbitrary square complex matrix is expressible as a product of two matrices of very special types, namely, positive semidefinite and unitary. The fact that our knowledge of these two types is extensive lends importance to this result. We shall take only the case of a nonsingular matrix (see Theorem 9–30).

THEOREM 9–29. *Let J be a positive semidefinite matrix. Then there is a unique positive semidefinite matrix H such that $H^2 = J$. Moreover, H has the same rank as J.*

Since J is Hermitian, hence similar to a diagonal matrix, it has a decomposition into principal idempotents (Theorem 9–5),

$$(37) \qquad J = j_1 E_1 + \cdots + j_t E_t$$

where every $j_i \geqq 0$ and every E_i is Hermitian. (Why?) If there exists a matrix H as described in the theorem, H also has such a decomposition,

$$H = h_1 F_1 + \cdots + h_q F_q, \quad h_i \geqq 0,$$
$$H^2 = h_1^2 F_1 + \cdots + h_q^2 F_q = J.$$

Since the last equation clearly is a principal idempotent decomposition for J, the uniqueness of this decomposition implies that the distinct numbers $h_1^2, \ldots, h_q^2$ are the same set as $j_1, \ldots, j_t$, and that $F_1, \ldots, F_q$ are the same, in some order, as $E_1, \ldots, E_t$. We then have $q = t$ and, changing the subscripts, if necessary,

$$h_i = \sqrt{j_i} \geqq 0, \quad F_i = E_i. \qquad (i = 1, \ldots, t)$$

Thus the only possible matrix H fulfilling the requirements of the theorem is the matrix

$$(38) \qquad H = \sqrt{j_1} E_1 + \cdots + \sqrt{j_t} E_t$$

obtained from (37) by replacing each characteristic root j_i by its nonnegative square root.

This construction always produces a matrix H such that $H^2 = J$. Moreover, $\bar{H}' = H$, so the constructed H is always Hermitian. Since (38) is a principal idempotent decomposition, the distinct characteristic roots of H must be $\sqrt{j_i}$ ($i = 1, \ldots, t$), and the fact that these are not negative implies (Theorem 9–25) that H is positive semidefinite.

THEOREM 9–30. *If A is a nonsingular complex matrix, there exist unique matrices H and U such that H is positive definite, U is unitary, and $A = HU$.*

The clue to the construction of H and U is found by assuming the result. Then

$$A\bar{A}' = HU\bar{U}'\bar{H}' = HUU^{-1}H = H^2.$$

Thus H is a positive definite "square root" of the positive definite matrix $J = A\bar{A}'$, and by Theorem 9–29 there is only one such matrix H. If then $A = HU = HU_1$, the nonsingularity of H implies that $U_1 = U$. This proves that if the matrices H and U in Theorem 9–30 exist, they are unique.

The way to prove the existence is now clear. The matrix $J = A\bar{A}'$ is positive definite and determines (Theorem 9–29) a positive definite matrix H such that $H^2 = J$. It remains only to show that $U = H^{-1}A$ is unitary. First observe that since H is positive definite, so is H^{-1}. (Why?) Then

$$U\bar{U}' = H^{-1}A\bar{A}'\overline{H^{-1}}' = H^{-1}H^2H^{-1} = I.$$

Thus U is unitary and the proof is complete.

It is well known that each nonzero complex number z is expressible uniquely in the polar form $z = hu$, where h is real and positive and u has the form

$$u = \cos t + i \sin t.$$

Since the inverse of u is its conjugate, the analogy with Theorem 9–30 is apparent and for this reason the factorization $A = HU$ in Theorem 9–30 is known as the *polar decomposition* of A. Recall that h in the expression $z = hu$ is the absolute value of z,

$$h^2 = z\bar{z},$$

so that h is the positive square root of $z\bar{z}$, just as H above is the "positive square root" of $A\bar{A}'$.

EXERCISES

1. Prove the modification of Theorem 9–30, which asserts that A is expressible uniquely as $A = UH$, U unitary, H positive definite.

★2. Prove that the nonsingular matrix A of Theorem 9–30 is normal if and only if $HU = UH$.

★3. Generalize Theorem 9–29 to the case $H^k = J$, k being any positive integer.

★4. Let B be positive semidefinite. Prove that if k is any positive integer, a unitary matrix commuting with B^k necessarily commutes with B.

5. Let A be a nonsingular normal matrix $A = HU$ as in Theorem 9–30. Prove that if k is a positive integer and A_1 is a normal matrix which is a kth root of A ($A_1^k = A$) then $A_1 = H_1U_1$, where H_1 is the unique positive definite matrix such that $H_1^k = H$ and U_1 is a unitary matrix such that $U_1^k = U$. Conversely, every such normal matrix $A_1 = H_1U_1$ is a kth root of A.

APPENDIX

9–16 Small vibrations. A dynamic system with n degrees of freedom is described in terms of n variables $q_1, \ldots, q_n$, which are called *generalized displacements.* These are functions of the time t whose values at each moment completely determine the configuration of the system at that moment. If the system is conservative and is undergoing small vibrations about a position of stable equilibrium, its potential energy U and kinetic energy T may, to a high degree of accuracy, be represented by quadratic forms

$$U = \tfrac{1}{2}Q'KQ, \quad T = \tfrac{1}{2}\dot{Q}'M\dot{Q},$$

where $K = (k_{ij})$ and $M = (m_{ij})$ are $n \times n$ symmetric matrices of constants,

$$Q = \mathrm{col}\,(q_1, \ldots, q_n), \quad \dot{Q} = \mathrm{col}\,(\dot{q}_1, \ldots, \dot{q}_n),$$

$$\dot{q}_i = \frac{dq_i}{dt}. \qquad\qquad (i = 1, \ldots, n)$$

Lagrange's equations for the vibrating system may be written as

$$\frac{d}{dt}\left(\frac{\partial T}{\partial \dot{q}_i}\right) + \frac{\partial U}{\partial q_i} = 0. \qquad (i = 1, \ldots, n)$$

Then

$$U = \tfrac{1}{2}\Sigma k_{ij}q_iq_j, \quad T = \tfrac{1}{2}\Sigma m_{ij}\dot{q}_i\dot{q}_j,$$

and the reader can verify that

$$\frac{\partial U}{\partial q_i} = \Sigma k_{ij}q_j, \quad \frac{\partial T}{\partial \dot{q}_i} = \Sigma m_{ij}\dot{q}_j.$$

Thus Lagrange's equations become

$$\sum_j m_{ij}\ddot{q}_j + \sum_j k_{ij}q_j = 0, \qquad (i = 1, \ldots, n)$$

where

$$\ddot{q}_j = \frac{d^2q_j}{dt^2}.$$

Many problems in physics and engineering depend on the solution of a system of equations of vibration like those above. In matrix form this system may be written as

$$(39) \qquad\qquad M\ddot{Q} + KQ = 0.$$

Since the kinetic energy is positive except when all of the generalized velocities $\dot{q}_j$ are zero, $M = (m_{ij})$ is a positive definite, real, symmetric matrix. The theory of pairs of quadratic forms (Theorem 9–13) may now be applied to (39). A nonsingular linear substitution

$$Q = PW, \quad P = (p_{ij}), \quad W = \text{col}\,(w_1, \ldots, w_n),$$

leads to the equations

$$\ddot{Q} = P\ddot{W}, \quad MP\ddot{W} + KPW = 0,$$
$$(P'MP)\ddot{W} + (P'KP)W = 0.$$

As we have seen in Theorem 9–13, it is possible to choose P so that

$$P'MP = I, \quad P'KP = \text{diag}\,(r_1, \ldots, r_n).$$

Then the system becomes

$$\ddot{W} + \text{diag}\,(r_1, \ldots, r_n)W = 0,$$

$$\frac{d^2 w_i}{dt^2} + r_i w_i = 0, \qquad\qquad (i = 1, \ldots, n)$$

so that the variables are separated, that is, each equation involves only one w_i. In theory, then, we have shown how to solve the equations of motion of a conservative dynamical system undergoing small oscillations about a position of stable equilibrium.

9–17 Iterative solution of frequency equation. The job of finding the characteristic roots of a matrix is very often the crux of a problem. One of the important instances occurs in the equations of vibrations,

$$(40) \qquad\qquad M\ddot{Q} + KQ = 0,$$

discussed in the preceding section. The solutions $Q = (q_1, \ldots, q_n)$ of primary interest are of the form

$$q_i = a_i \sin rt, \qquad\qquad (i = 1, \ldots, n)$$

where the a_i are constants (called amplitudes) which are not all zero, and r is a nonzero constant. If we confine attention to such solutions and write

$$\alpha = \text{col}\,(a_1, \ldots, a_n),$$

then
$$Q = \alpha \sin rt, \quad \ddot{Q} = -r^2 \sin rt \cdot \alpha,$$
and the equation (40) becomes
$$(-r^2 \sin rt \, M + \sin rt \, K)\alpha = 0,$$
or

(41) $(r^2 I - A)\alpha = 0, \quad A = M^{-1}K.$

There exists a nonzero vector α satisfying (41) if and only if

(42) $|r^2 I - A| = 0.$

Thus r^2 must be a characteristic root of A, and α a characteristic vector of A corresponding to r^2. Because r is an angular frequency of vibration, (42) is known in engineering circles as the *frequency equation*.

For matrices A of large size it may be difficult even to calculate the characteristic polynomial of A, let alone its roots. There is a popular method for approximating these roots and the corresponding vectors α, which is known as the method of *iteration*, since it involves repeated multiplication by A. A brief exposition of this method will now be given.

Let A be a nonsingular complex matrix. The notation for its distinct characteristic roots $s_1, \ldots, s_t$ may be so chosen that

$$|s_1| \geqq |s_2| \geqq \cdots \geqq |s_t| > 0.$$

We now make two assumptions: (a) A is similar to a diagonal matrix; (b) $|s_1| > |s_2|$.

By the first assumption A has a principal idempotent decomposition

$$A = s_1 E_1 + \cdots + s_t E_t,$$
and for each positive integer k
$$A^k = s_1^k E_1 + \cdots + s_t^k E_t.$$

By the second assumption, if k is large the quantities s_i^k, $i > 1$, are negligible in comparison with s_1^k. Stated differently, the matrix

$$s_1^{-k} A^k = E_1 + (s_2/s_1)^k E_2 + \cdots + (s_t/s_1)^k E_t$$

approaches E_1 as a limit when $k \to \infty$. That is, each element of $s_1^{-k} A^k$ is a function of k which, as k increases, approaches the correspondingly placed element of E_1. For large k we may therefore write

$$s_1^{-k} A^k \doteq E_1,$$

where $\doteq$ denotes approximate equality. Then

$$A^{k+1} \doteq s_1^{k+1}E_1 = s_1(s_1^k E_1),$$
$$A^{k+1} \doteq s_1 A^k.$$

The last equation reveals the method for approximating s_1. We compute successive powers of A until we reach a power A^{k+1} which is nearly equal to a scalar multiple of the preceding power, A^k. The scalar is the approximate value of s_1, the largest characteristic root of A, and the approximation may be improved by continuing to higher powers.

The following modification not only simplifies the computation, but also leads to an approximate characteristic vector α corresponding to s_1.

Let α_0 be any nonzero column vector, so that $A\alpha_0 \neq 0$. Define

$$\alpha_k = A^k \alpha_0. \qquad (k = 1, 2, \ldots)$$

Thus

$$\alpha_k = (s_1^k E_1 + \cdots + s_t^k E_t)\alpha_0$$
$$= s_1^k(E_1\alpha_0) + \cdots + s_t^k(E_t\alpha_0).$$

As in the preceding discussion, $s_1^{-k}\alpha_k$ approaches $E_1\alpha_0$ as a limit, so for large k

$$\alpha_k \doteq s_1^k E_1\alpha_0, \quad \alpha_{k+1} \doteq s_1^{k+1}E_1\alpha_0,$$
$$\alpha_{k+1} \doteq s_1\alpha_k.$$

Thus we compute the successive products

$$\alpha_1 = A\alpha_0, \quad \alpha_2 = A\alpha_1, \ldots$$

until we find a vector $\alpha_{k+1} = A\alpha_k$ which is approximately a scalar multiple of the preceding vector α_k. The scalar factor is the approximate value of s_1, an approximation which may be improved by continuing the process.

Since the "equation" $\alpha_{k+1} \doteq s_1\alpha_k$ may be written as $A\alpha_k \doteq s_1\alpha_k$, it can be shown that the vectors α_k approximate a characteristic vector of A corresponding to s_1. In numerical work it often happens, however, that the coordinates of α_k increase without bound as k increases. In fact this happens whenever $|s_1| > 1$. This nuisance may be avoided by the following (final!) modification of the process for approximating both s_1 and a characteristic vector corresponding to s_1.

After $A\alpha_i$ is computed, the coordinate t_i in some fixed position is made 1 by removing a factor t_i from all the coordinates of $A\alpha_i$.

The vector so constructed, rather than $A\alpha_i$ itself, is defined to be α_{i+1}, and this is done for $i = 0, 1, 2, \ldots$. The discussion above then implies that the factors $t_0, t_1, \ldots$, tend towards s_1.

EXAMPLE:

$$A = \begin{bmatrix} 1 & 2 \\ 2 & 2 \end{bmatrix}, \quad \alpha_0 = \begin{bmatrix} 1 \\ 0 \end{bmatrix},$$

$$A\alpha_0 = \begin{bmatrix} 1 \\ 2 \end{bmatrix}, \quad t_0 = 1, \quad \alpha_1 = \begin{bmatrix} 1 \\ 2 \end{bmatrix},$$

$$A\alpha_1 = \begin{bmatrix} 5 \\ 6 \end{bmatrix}, \quad t_1 = 5, \quad \alpha_2 = \begin{bmatrix} 1 \\ \frac{6}{5} \end{bmatrix},$$

$$A\alpha_2 = \begin{bmatrix} \frac{17}{5} \\ \frac{22}{5} \end{bmatrix}, \quad t_2 = \frac{17}{5} = 3.40, \quad \alpha_3 = \begin{bmatrix} 1 \\ \frac{22}{17} \end{bmatrix},$$

$$A\alpha_3 = \begin{bmatrix} \frac{61}{17} \\ \frac{78}{17} \end{bmatrix}, \quad t_3 = \frac{61}{17} = 3.59, \quad \alpha_4 = \begin{bmatrix} 1 \\ \frac{78}{61} \end{bmatrix},$$

$$A\alpha_4 = \begin{bmatrix} \frac{217}{61} \\ \frac{278}{61} \end{bmatrix}, \quad t_4 = \frac{217}{61} = 3.56.$$

In this simple case the characteristic equation is $x^2 - 3x - 2 = 0$ and its roots are $x = 3.56$, $x = -0.56$.

The physical facts of various practical problems give hints as to the coordinates of α. When these hints are used in choosing the initial vector α_0, the convergence of the iteration process may be hastened.

For some purposes only the dominant root is required, or sometimes only the root s_n which has smallest absolute value. Since s_n^{-1} is the dominant root of A^{-1}, it may be found by applying the iteration procedure to this matrix.

For procedures and proofs when the hypotheses are not valid see Reference 8, which also presents methods of computing the roots other than s_1 and s_n.

9–18 Orthogonal matrices. In the language of automorphs, orthogonal matrices may be defined as the real, congruent automorphs of identity matrices. Taking $S = I$ in Theorem 5–24 then gives all orthogonal matrices P such that $I + P$ is nonsingular. The latter condition is equivalent to the property that -1 is not a characteristic root of P. (Why?) Orthogonal matrices with this property are called *proper*.

The hypotheses that $S + K$ and $S - K$ shall be nonsingular are always satisfied in the present case in which $S = I$.   (Why?)   In summary we have proved

THEOREM 9-31.   *Let K be any real, skew matrix.   Then the matrix*

$$P = (I + K)^{-1}(I - K)$$

is properly orthogonal, and every properly orthogonal matrix P is expressible in this way.

We thus have a construction for proper orthogonal matrices in terms of a type of matrix (skew) which we know how to construct.

The last theorem leads to an interesting analogy between properly orthogonal matrices and complex numbers having unit norm.   The norm of $a + bi$ is defined to be

$$(a + bi)(a - bi) = a^2 + b^2,$$

that is, the square of the absolute value of $a + bi$.   Then $a + bi$ has unit norm if and only if $a^2 + b^2 = 1$.   In order to develop this analogy we first consider an analogy between arbitrary square matrices and arbitrary complex numbers.

Let A be a square matrix over $\Re$.   Then A is uniquely expressible as

$$A = S + K, \quad S \text{ symmetric, } K \text{ skew,}$$

whence

$$A' = S - K.$$

For any complex number

(43) $$z = a + bi$$

the conjugate is

(44) $$\bar{z} = a - bi.$$

Comparison of (44) and (43) with the two preceding equations reveals a clear analogy:

$$A = S + K \to z = a + bi,$$
$$S \to a,$$
$$K \to bi,$$
$$A' = S - K \to \bar{z} = a - bi.$$

That is, the transpose process for matrices is analogous to the conjugate process for complex numbers; symmetric matrices are those which equal their transposes, just as real numbers equal their conjugates; skew matrices are the negatives of their transposes just as purely imaginary numbers are the negatives of their conjugates.

Since an orthogonal matrix is one whose transpose is its inverse, the above analogy would associate this class of matrices with the class of complex numbers z such that

$$\bar{z} = z^{-1}.$$

These are exactly the numbers z having unit norm. Moreover, properly orthogonal matrices would then be associated with those numbers z which are $\neq -1$ and have unit norm.

We may show that a complex number $z \neq -1$ has unit norm if and only if z is expressible in the form

$$(45) \qquad z = (1 + ci)^{-1}(1 - ci),$$

where ci is purely imaginary or zero.

If (45) holds, then

$$\bar{z} = (1 - ci)^{-1}(1 + ci),$$

so that $z\bar{z} = 1$. Conversely, let $z\bar{z} = 1$. Any square root of z may be denoted by

$$p - qi.$$

Then $p - qi$ also has unit norm,

$$(p + qi)(p - qi) = 1,$$
$$(p + qi)^{-1} = p - qi,$$
$$z = (p - qi)(p - qi)$$
$$= (p + qi)^{-1}(p - qi)$$
$$= (1 + ci)^{-1}(1 - ci),$$

where $c = q/p$. This completes the proof that the set of all complex numbers of unit norm coincides with the set of numbers (45). (If $p = 0$ the construction fails, but then one can show that $z = -1$, an excluded case.)

Since skew matrices K correspond, in our analogy, to pure imaginaries ci, the formula in Theorem 9-31 is clearly the analogue of (45). For further extensions of these ideas see Reference 7, pp. 162-167, and Reference 9, p. 122.

9-19 Commutativity of Hermitian matrices. Questions of commutativity of matrices are of increasing importance in the applications of matrices. An interesting sample of a commutativity theorem follows.

THEOREM 9–32. *Let A and B be Hermitian matrices. Then A commutes with B if and only if A and B are polynomials, with real coefficients, in a common Hermitian matrix C.*

If A and B are polynomials in a common matrix C, it is evident that A and B commute (even if the coefficients of the polynomials are not real). Conversely, A and B have decompositions into principal idempotents,

$$A = a_1E_1 + \cdots + a_rE_r,$$
$$B = b_1F_1 + \cdots + b_sF_s,$$

where the idempotents E_j and F_k are Hermitian. Since $AB = BA$ it follows (Theorem 9–5) that B commutes with each E_j. From a different viewpoint, since E_j commutes with B, E_j must commute with each principal idempotent of B. In short

$$E_jF_k = F_kE_j$$

for all j and k.

Consider a matrix

$$C = \sum_{j,k} c_{jk}E_jF_k,$$

where the rs real numbers c_{jk} are all distinct from one another. The proof that C is Hermitian is trivial. By Exercise 9, Section 4–7, we can pick polynomials $e(x)$ and $f(x)$ in $\mathfrak{R}[x]$ such that

$$e(c_{jk}) = a_j, \quad f(c_{jk}) = b_k$$

for all j and k. We then find, since E_j commutes with F_k, that

$$C^2 = (\Sigma c_{jk}E_jF_k)(\Sigma c_{jk}E_jF_k)$$
$$= \Sigma c_{jk}^2 E_jF_k,$$
$$C^n = \Sigma c_{jk}^n E_jF_k,$$

whence

$$e(C) = \sum_{j,k} a_jE_jF_k$$
$$= \sum_k \left(\sum_j a_jE_j\right)F_k$$
$$= \sum_k AF_k = A\sum_k F_k = A.$$

Similarly,

$$f(C) = \sum_{j,k} b_kE_jF_k$$
$$= \sum_j \sum_k E_j(b_kF_k) = B.$$

This completes the proof of the theorem.

An application of this result to the unitary reduction of a pair of Hermitian matrices is obtainable immediately.

THEOREM 9–33.   *Let A and B be n × n Hermitian matrices.   Then there exists a unitary matrix P such that both $P^{-1}AP$ and $P^{-1}BP$ are diagonal, if and only if A commutes with B.*

Suppose that the matrix P exists.   Since two diagonal matrices commute,

$$P^{-1}APP^{-1}BP = P^{-1}BPP^{-1}AP,$$
$$P^{-1}ABP = P^{-1}BAP,$$
$$AB = BA.$$

Conversely, if $AB = BA$, the previous theorem applies so that

$$A = e(C), \quad B = f(C),$$

where C is Hermitian and $e(x)$ and $f(x)$ are polynomials.   There is a unitary matrix P such that $P^{-1}CP$ is a diagonal matrix D, whence $e(D)$ is also diagonal.   But

$$P^{-1}AP = e(P^{-1}CP) = e(D),$$
$$P^{-1}BP = f(P^{-1}CP) = f(D),$$

so that $P^{-1}AP$ and $P^{-1}BP$ are both diagonal.

CHAPTER 10

LINEAR TRANSFORMATIONS

10–1 Coordinates and bases. Let

(1)
$$\alpha_1, \ldots, \alpha_n$$

form a basis of $V_n(\mathfrak{F})$. Then for each vector

(2)
$$\xi = \text{col } (c_1, \ldots, c_n)$$

of $V_n(\mathfrak{F})$ there are scalars $a_1, \ldots, a_n$ such that

(3)
$$\xi = a_1\alpha_1 + \cdots + a_n\alpha_n.$$

The scalars a_i are uniquely determined by ξ and the basis (1), and are called *coordinates* of ξ relative to this basis. To express this relationship briefly we write

(4)
$$\xi_\alpha = \text{col } (a_1, \ldots, a_n).$$

If the particular basis (1) is clearly understood, (4) determines the vector ξ just as well as (2).

In $V_2(\mathfrak{R})$, for example, consider the basis

(5)
$$\alpha_1 = \begin{bmatrix} 1 \\ 1 \end{bmatrix}, \quad \alpha_2 = \begin{bmatrix} 1 \\ -1 \end{bmatrix}.$$

For a particular ξ we shall find ξ_α:

$$\xi = \begin{bmatrix} 2 \\ 3 \end{bmatrix}, \quad \xi_\alpha = \begin{bmatrix} a_1 \\ a_2 \end{bmatrix},$$

$$\begin{bmatrix} 2 \\ 3 \end{bmatrix} = a_1\alpha_1 + a_2\alpha_2 = \begin{bmatrix} a_1 + a_2 \\ a_1 - a_2 \end{bmatrix},$$

$$\left. \begin{array}{l} a_1 + a_2 = 2 \\ a_1 - a_2 = 3 \end{array} \right\}, \quad a_1 = \tfrac{5}{2}, \ a_2 = -\tfrac{1}{2},$$

(6)
$$\xi_\alpha = \begin{bmatrix} \tfrac{5}{2} \\ -\tfrac{1}{2} \end{bmatrix}.$$

Conversely, if coordinates η_α of a vector η are given, η is determined. For example,

$$\eta_\alpha = \begin{bmatrix} 3 \\ -1 \end{bmatrix}, \quad \eta = 3\alpha_1 - \alpha_2 = \begin{bmatrix} 2 \\ 4 \end{bmatrix}.$$

If the basis (1) of $V_n(\mathfrak{F})$ is the basis of unit vectors, for each vector ξ we have

(7) $\xi = \mathrm{col}\ (c_1,\ \dots,\ c_n) = \xi_\alpha.$

The basis of unit vectors is the only one for which the identity (7) is valid. However, coordinates relative to an arbitrary basis are about as handy as those relative to unit vectors. Sums and scalar products can be calculated in terms of coordinates relative to any basis:

(8) $\xi_\alpha = \mathrm{col}\ (a_1,\ \dots,\ a_n),\quad \eta_\alpha = \mathrm{col}\ (b_1,\ \dots,\ b_n)$

imply

(9)
$$(\xi + \eta)_\alpha = \mathrm{col}\ (a_1 + b_1,\ \dots,\ a_n + b_n),$$
$$(k\xi)_\alpha = \mathrm{col}\ (ka_1,\ \dots,\ ka_n) = k\xi_\alpha.$$

The first of these equations (9) means that

$$\xi + \eta = \Sigma(a_i + b_i)\alpha_i,$$

the truth of which is an obvious consequence of the meaning of (8). The second part of (9) is equally easy to prove.

THEOREM 10–1. *Let ξ, η, and ζ lie in $V_n(\mathfrak{F})$, and k lie in $\mathfrak{F}$. Then $\zeta = \xi + \eta$ if and only if $\zeta_\alpha = \xi_\alpha + \eta_\alpha$, and $\zeta = k\xi$ if and only if $\zeta_\alpha = k \cdot \xi_\alpha$.*

We have proved above that if $\zeta = \xi + \eta$, then $\zeta_\alpha = \xi_\alpha + \eta_\alpha$. Suppose now that the latter equation is true, whence $\zeta_\alpha = (\xi + \eta)_\alpha$ by (9). Since $\xi + \eta$ thus has the same n-tuple of coordinates as ζ, $\xi + \eta$ is ζ. The final property of Theorem 10–1 has a similar proof.

COROLLARY 10–1. *$\zeta = a\xi + b\eta$ if and only if $\zeta_\alpha = a \cdot \xi_\alpha + b \cdot \eta_\alpha$.*

The proof is left for the reader.

For $V_2(\mathfrak{R})$ and $V_3(\mathfrak{R})$ we know that coordinates of ξ relative to the basis of unit vectors are ordinary rectangular coordinates (of the end point of ξ). This fact may be generalized.

THEOREM 10–2. *For $V_2(\mathfrak{R})$ and $V_3(\mathfrak{R})$ coordinates relative to any orthonormal basis are ordinary rectangular coordinates.*

If we discuss $V_2(\mathfrak{R})$ the reader will be able to make his own proof for three dimensions. Let α_1, α_2 form the orthonormal basis, and let

$$\xi = a_1\alpha_1 + a_2\alpha_2.$$

As we saw in Section 9–4, α_1 and α_2 are perpendicular line segments of unit length. Hence there is a unique rectangular coordinate system such that α_1 and α_2 are the unit lengths along the positive directions on the x and y axes, respectively. Each vector $a_i\alpha_i$ has length $|a_i|$:

$$[(a_i\alpha_i)'(a_i\alpha_i)]^{\frac{1}{2}} = [a_i^2\alpha_i'\alpha_i]^{\frac{1}{2}} = |a_i|.$$

Since vectors add by the parallelogram law (Section 2–1), and since in the present case the parallelogram is a rectangle (see Figure 10–1) it is clear that the end point of ξ has coordinates a_1 and a_2.

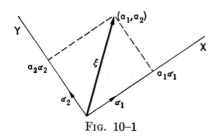

Rectangular coordinates in $V_n(\Re)$, for arbitrary n, may be defined simply as coordinates relative to any chosen orthonormal basis; and each orthonormal basis may be called a *rectangular coordinate system*.

Fig. 10–1

10–2 Coordinates relative to two bases. We frequently have occasion to change the basis of $V_n(\mathfrak{F})$. Suppose that we have the basis (1) and a new basis,

$$(10) \qquad\qquad \beta_1, \ldots, \beta_n.$$

Then the vector ξ in (2) is a linear combination of the β_i,

$$(11) \qquad\qquad \xi = b_1\beta_1 + \cdots + b_n\beta_n,$$

so that relative to the new basis ξ has coordinates

$$(12) \qquad\qquad \xi_\beta = \text{col } (b_1, \ldots, b_n).$$

It is not hard to get a general formula for the new coordinates (12) of each vector in terms of the old coordinates (4) and the matrix relating the new and old bases of $V_n(\mathfrak{F})$. Let M and N be the matrices whose ith columns are α_i and β_i, respectively, $i = 1, \ldots, n$. Then M and N are nonsingular, and we let $P = N^{-1}M$. The formulas (3) and (11) expressing the same vector ξ in terms of the two bases may now be written as matrix products

$$\xi = N\xi_\beta = M\xi_\alpha,$$

so that

$$(13) \qquad\qquad \xi_\beta = P\xi_\alpha.$$

THEOREM 10–3. *If a vector ξ of $V_n(\mathfrak{F})$ has coordinates $\xi_\alpha = \text{col } (a_1, \ldots, a_n)$ relative to one basis and coordinates $\xi_\beta = \text{col } (b_1, \ldots, b_n)$ relative to another basis, the two sets of coordinates are related by* *equation* (13) *in which P is a nonsingular matrix independent of ξ and determined by the two bases.*

Note (by Section 3–13) that an arbitrary nonsingular P can be made to occur in (13) by appropriate choice of the new basis.

EXERCISES

1. If the old and new bases in Theorem 10–3 are orthonormal, show that P is orthogonal. Conversely, assume that P is orthogonal and prove that the new basis is orthonormal if and only if the old basis is.

2. In $V_2(\mathfrak{R})$ consider the basis (5) and the new basis

$$\beta_1 = \begin{bmatrix} 1 \\ 2 \end{bmatrix}, \quad \beta_2 = \begin{bmatrix} 3 \\ 1 \end{bmatrix}.$$

Find the matrix P in (13) for this case and compute the coordinates ξ_β of $\xi = \mathrm{col}\,(2, 3)$ from P and (6). Check your answer.

3. Prove Corollary 10–1.

10–3 Linear transformations. In analytic geometry a curve in the xy plane is said to be symmetric with respect to the x axis if for each point (x,y) on the curve, the point $(x,-y)$ also lies on the curve. Stated otherwise, the correspondence

(14) $(x,y) \rightarrow (x,-y)$

(the arrow is read "goes to" or "corresponds to") carries each point of the curve into a point of the curve. Notice, however, that (14)

may be applied to every point (x,y) of the plane. Then (14) carries each point into its "mirror image" in the x axis, whence this correspondence of the xy plane with itself is called a *reflection* in the x axis. A curve is symmetric in the x axis if the reflection of the xy plane in this axis carries each point of the curve into a point of the curve.

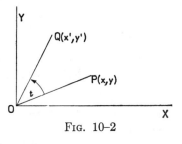

FIG. 10–2

This is but one instance of a correspondence of a set with itself. Another is the familiar example of rotation of the xy plane.* Here

_____

* The x and y axes are to remain fixed, and each point P with coordinates (x, y) is to move to a new position Q with coordinates (x', y') by rotating the line OP to a position OQ; all lines OP rotate through the same angle.

each point (x,y) is carried into a new point, or position, (x',y'), where

$$x' = x \cos t - y \sin t,$$
$$y' = x \sin t + y \cos t.$$

Correspondences of a set, in particular of a vector space, with itself occur so often in mathematics that they are known by many names: *correspondences, mappings, transformations, functions.* In $V_n(\mathfrak{F})$ we may consider a mapping T carrying each vector ξ into a well-defined vector, which we shall indicate by $T(\xi)$. The latter vector is called the *map* or *image* or *transform* of ξ under T. If $\xi = $ col $(c_1, \ldots, c_n)$, various mappings may carry ξ into

$$T(\xi) = \text{col } (c_1^2, c_2^2, \ldots, c_n^2),$$
$$T(\xi) = \text{col } (-c_1, -c_2, \ldots, -c_n),$$
$$T(\xi) = \text{col } (c_2, c_3, \ldots, c_n, -c_1),$$
$$T(\xi) = \text{col } (c_1 + c_2, c_2 + c_3, \ldots, c_n + c_1).$$

In these examples $T(\xi)$ is defined by formulas in terms of the coordinates $c_1, \ldots, c_n$ of ξ relative to the unit vectors. We can equally well use coordinates relative to any other basis. However, a transformation may be defined without any algebraic formulas at all. Moreover, for any given T, the vector $T(\xi)$ is completely determined by ξ; it does not depend on any particular basis of the vector space.

DEFINITION 1.* A *linear transformation* of $V_n(\mathfrak{F})$ is a mapping $\xi \to T(\xi)$ of $V_n(\mathfrak{F})$ into itself such that

$$T(a_1\xi_1 + a_2\xi_2) = a_1T(\xi_1) + a_2T(\xi_2)$$

for every pair of vectors ξ_1 and ξ_2 of $V_n(\mathfrak{F})$ and every pair of scalars a_1 and a_2.

Linear transformations are of great importance in mathematics. As we shall see, the theory of square matrices amounts to a particular way of studying linear transformations of vector spaces.

THEOREM 10–4. *A transformation T of $V_n(\mathfrak{F})$ is linear if and only if*

$$T(\xi_1 + \xi_2) = T(\xi_1) + T(\xi_2), \quad T(a\xi) = aT(\xi)$$

for every scalar a and all vectors ξ, ξ_1, and ξ_2.

---

* Linear transformations appeared concretely in Chapter 9. It is easy to verify that those transformations have the property stated in the present abstract definition. But this verification will also be immediately apparent from the next section.

The proof is an exercise for the reader. In the four examples above of transformations T, only the last three are linear. The fact that these are the only cases in which the formulas are linear in terms of the c_i is far from a coincidence, as we shall proceed to show.

EXERCISES

1. Prove Theorem 10–4.

2. If T is a linear transformation of $V_n(\mathfrak{F})$, prove that $T(a_1\xi_1 + \cdots + a_r\xi_r) = a_1T(\xi_1) + \cdots + a_rT(\xi_r)$ for every positive integer r, all scalars a_i, and all vectors ξ_i in $V_n(\mathfrak{F})$.

3. Prove that the correspondence (14) on $V_2(\mathfrak{R})$ is a linear transformation.

4. Let $\alpha_1, \ldots, \alpha_n$ be a basis of $V_n(\mathfrak{F})$, and $\beta_1, \ldots, \beta_n$ any set of n vectors in the same space. Show that there is one and only one linear transformation T of $V_n(\mathfrak{F})$ such that $T(\alpha_i) = \beta_i$, $i = 1, \ldots, n$.

5. Let T be a linear transformation of $V_n(\mathfrak{F})$. Let M be the totality of vectors ξ in $V_n(\mathfrak{F})$ such that $T(\xi) = \xi$, and N the totality of vectors ξ in $V_n(\mathfrak{F})$ such that $T(\xi) = 0$. Prove that M and N are subspaces.

6. Prove that the subspaces M and N of Exercise 5 have only the zero vector in common.

7. Show that the following transformations T are linear and have the property $T[T(\xi)] = T(\xi)$ for every ξ in $V_n(\mathfrak{F})$.
   (a) $T(\xi) = \xi$ for all ξ.
   (b) $T(\xi) = 0$ for all ξ.
   (c) If $\xi = \text{col}\ (x_1, \ldots, x_n)$ and k is a fixed positive integer $\leq n$, then $T(\xi) = \text{col}\ (x_1, \ldots, x_k, 0, \ldots, 0)$.

8. Suppose that the linear transformation T in Exercise 5 has the property that $T[T(\xi)] = T(\xi)$ for all ξ in $V_n(\mathfrak{F})$, that is, applying the transformation twice has the same effect as applying it once. Prove that (a) the sum of the subspaces M and N is $V_n(\mathfrak{F})$; (b) the sum of the dimensions of M and N is n.

10–4 Matrix representation. Let

$$(15) \qquad \alpha_1, \ldots, \alpha_n$$

be a basis of $V_n(\mathfrak{F})$ and T a linear transformation. Then the vectors

$$T(\alpha_1), \ldots, T(\alpha_n),$$

like any other vectors of $V_n(\mathfrak{F})$, are expressible in terms of the basis (15), so that

$$(16) \qquad T(\alpha_j) = \sum_{i=1}^{n} \alpha_i t_{ij}, \qquad (i = 1, \ldots, n)$$

the coefficients t_{ij} being written on the right for later convenience. In terms of the matrices whose column vectors are the $T(\alpha_j)$ and the α_i, respectively, (16) becomes

$$(17) \qquad [T(\alpha_1), \ldots, T(\alpha_n)] = (\alpha_1, \ldots, \alpha_n)T_0,$$
$$T_0 = (t_{ij}).$$

The utility of the matrix T_0 extends much further than is indicated by (17), where T_0 is employed in computing the images $T(\alpha_j)$ of the basis vectors.

Suppose that $\xi_\alpha = \text{col}(a_1, \ldots, a_n)$ relative to the basis (15). Then

$$(18) \qquad \xi = a_1\alpha_1 + \cdots + a_n\alpha_n,$$
$$T(\xi) = a_1T(\alpha_1) + \cdots + a_nT(\alpha_n) = \eta,$$
$$(19) \qquad T(\xi) = [T(\alpha_1), \ldots, T(\alpha_n)]\xi_\alpha$$
$$= (\alpha_1, \ldots, \alpha_n)T_0\xi_\alpha.$$

Thus the unique coordinates η_α of $\eta = T(\xi)$ must be

$$(20) \qquad \eta_\alpha = T_0\xi_\alpha.$$

In summary:

THEOREM 10–5. *Let T be a linear transformation of $V_n(\mathfrak{F})$ mapping ξ into $\eta = T(\xi)$. Then, relative to some fixed basis (15) the coordinates η_α of η are computed from the coordinates ξ_α of ξ by (20). The matrix $T_0 = (t_{ij})$ is completely determined by the effect (16) of T on the given basis.*

In brief, multiplication of the column of coordinates of any vector ξ on the left by the square matrix T_0 produces the column of coordinates of $T(\xi)$. Because of this important relationship between the linear transformation T on the one hand, and the matrix T_0 appearing in (17) and (20) on the other, we call T_0 the *matrix representation* of T relative to the basis (15).

The resemblance between (20) and (13) is apparent — a column is computed by multiplying a column by a square matrix. The distinction in content, however, is very sharp. In (13) we contemplate a single vector ξ and two bases of $V_n(\mathfrak{F})$. The coordinates of ξ relative to the two bases are connected by (13). In (20) we contemplate two vectors and a single basis. The coordinates of the two vectors relative to the same basis are connected by (20).

A further distinction is that the matrix P in (13) is always non-singular. The matrix T_0 in (20) may be singular. In fact the next result asserts that T_0 may be an arbitrary $n \times n$ matrix over $\mathfrak{F}$.

THEOREM 10–6. *Let T_0 be any $n \times n$ matrix over $\mathfrak{F}$ and let (15) be a basis of $V_n(\mathfrak{F})$. Then there is one and only one linear transformation T of $V_n(\mathfrak{F})$ which is represented relative to (15) by T_0.*

Each vector ξ has coordinates ξ_α relative to the chosen basis (15). Then as in (20) the product $T_0\xi_\alpha$ is a well-defined n-tuple η_α of coordinates determining a vector η. Thus we have defined a transformation

$$\xi \to \eta = T(\xi),$$

which we shall prove linear. If

$$\xi = a_1\xi_1 + a_2\xi_2,$$

we must prove that

$$T(\xi) = a_1 T(\xi_1) + a_2 T(\xi_2).$$

By Corollary 10–1 it suffices to prove that

$$(21) \qquad T(\xi)_\alpha = a_1 T(\xi_1)_\alpha + a_2 T(\xi_2)_\alpha.$$

The same corollary implies that

$$\xi_\alpha = a_1\xi_{1\alpha} + a_2\xi_{2\alpha},$$

whence

$$(22) \qquad T_0\xi_\alpha = a_1 T_0\xi_{1\alpha} + a_2 T_0\xi_{2\alpha}.$$

By definition of the transformation T,

$$T_0\xi_\alpha = T(\xi)_\alpha, \qquad T_0\xi_{i\alpha} = T(\xi_i)_\alpha,$$

so that (22) is equivalent to (21), and T is linear. By virtue of (20) the transformation T constructed in this proof is the only possible one having T_0 as its matrix representation relative to (15). This completes the proof.

By means of this result each $n \times n$ matrix C over $\mathfrak{F}$ may be interpreted geometrically in a space of n dimensions. Select any basis $\alpha_1, \ldots, \alpha_n$ of $V_n(\mathfrak{F})$. Then C determines and represents a linear transformation

$$\xi_\alpha \to C\xi_\alpha.$$

If C happens to be nonsingular, it may be given this and also a second interpretation. For C also determines a new basis $\beta_1, \ldots, \beta_n$, and the vector

$$C\xi_\alpha = \xi_\beta$$

is the n-tuple of coordinates of ξ relative to the new basis.

This dual interpretation appears also in analytic geometry in connection with the equations

$$x' = x \cos t - y \sin t,$$
$$y' = x \sin t + y \cos t.$$

These may be interpreted (see Figure 10-2) as a rotation of all vectors OP through a common angle t. In this sense the equations

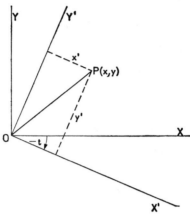

FIG. 10-3

define a linear transformation col $(x,y) \to$ col (x',y'). But there is the alternative meaning (Figure 10-3), in which the vectors OP, or points P, do not move: The coordinate axes are rotated through an angle $-t$, so that each point (x,y) has new coordinates (x',y').

EXERCISES

1. In the xy plane each vector is first reflected in the y axis, then doubled in length. Find the matrix representing this linear transformation relative to the basis of unit vectors.

2. Find the matrix representing a linear transformation T of $V_2(\Re)$ relative to the basis $\alpha_1 = $ col $(1,\ 1)$, $\alpha_2 = $ col $(1,\ -1)$, if for $\xi = $ col $(1,\ 2)$ and $\eta = $ col $(2,\ 1)$ we have $T(\xi) = $ col $(1,\ 0)$ and $T(\eta) = $ col $(2,\ 3)$.

★3. If T is a linear transformation of $V_n(\mathfrak{F})$ and V is a subspace, prove that the totality of vectors $T(\xi)$, ξ varying over V, forms a subspace; that is, a linear transformation of $V_n(\mathfrak{F})$ maps subspaces into subspaces.

10-5 Change of basis. The matrix A representing a linear transformation T is determined only when a basis of $V_n(\mathfrak{F})$ is selected. Relative to a new basis, T will be represented by a new matrix B. The relation between B and A is investigated in this section.

Consider, for example, the linear transformation T of $V_2(\Re)$ defined by the matrix

$$A = \begin{bmatrix} \frac{1}{2} & \frac{1}{2}\sqrt{3} \\ \frac{1}{2}\sqrt{3} & -\frac{1}{2} \end{bmatrix}$$

relative to the basis of unit vectors u_1, u_2. Then every vector col (a_1, a_2) also has coordinates a_1 and a_2, and

$$\xi = \begin{bmatrix} a_1 \\ a_2 \end{bmatrix}, \quad T(\xi) = \begin{bmatrix} a_1' \\ a_2' \end{bmatrix} = A \begin{bmatrix} a_1 \\ a_2 \end{bmatrix},$$

so that

(23)
$$a_1' = \tfrac{1}{2}(a_1 + \sqrt{3}a_2),$$
$$a_2' = \tfrac{1}{2}(\sqrt{3}a_1 - a_2).$$

Suppose now that we consider a new orthonormal basis

(24) $\beta_1 = $ col $(\tfrac{1}{2}\sqrt{3}, \tfrac{1}{2})$, $\beta_2 = $ col $(-\tfrac{1}{2}, \tfrac{1}{2}\sqrt{3})$.

Then the computations

$$T(\beta_i) = A\beta_i \qquad\qquad (i = 1, 2)$$

show that

$$T(\beta_1) = \beta_1,$$
$$T(\beta_2) = -\beta_2.$$

Since these equations display the expression for the $T(\beta_i)$ as linear combinations of β_1 and β_2, they show that the matrix representing T relative to the new basis (24) is

$$B = \begin{bmatrix} 1 & 0 \\ 0 & -1 \end{bmatrix}.$$

If a vector ξ has new coordinates $\xi_\beta = $ col (b_1, b_2), $T(\xi)$ then has new coordinates $B\xi_\beta = $ col $(b_1, -b_2)$. In terms of coordinates relative to the new basis, T is simply the transformation

(25) $$\begin{bmatrix} b_1 \\ b_2 \end{bmatrix} \rightarrow \begin{bmatrix} b_1 \\ -b_2 \end{bmatrix}.$$

Since the basis (24) is orthonormal, b_1 and b_2 are ordinary rectangular coordinates, and (25) shows that T is simply a reflection in the "β_1 axis." The fact that T is a reflection is not at all obvious from the original formulas (23) in terms of the matrix A, but is easily observed from the new formula (25) in terms of B. This brings out the importance of changing bases as a device for understanding the nature of a given linear transformation.

THEOREM 10-7. *Let T be a linear transformation of $V_r(\mathfrak{F})$ represented by a matrix A relative to a basis $\alpha_1, \ldots, \alpha_n$, and represented by a matrix B relative to a basis $\beta_1, \ldots, \beta_n$. Then B and A are similar,*

$$(26) \qquad A = P^{-1}BP,$$

where the nonsingular matrix P over $\mathfrak{F}$ is determined by the two bases (and is the matrix P of Theorem 10-3).

In short, similar matrices represent the same linear transformation T relative to two bases. This is perhaps the major motivation for the study of similarity. In every concrete application of similarity one may expect to find an underlying vector space on which the given similar matrices represent a common linear transformation. To prove (26) let ξ have coordinates ξ_α and ξ_β relative to the two bases. Likewise, $\eta = T(\xi)$ will have coordinates η_α and η_β, respectively, relative to the basis of α_i and the basis of β_i. By Theorem 10-3,

$$(27) \qquad \xi_\beta = P\xi_\alpha, \quad \eta_\beta = P\eta_\alpha.$$

The meaning of the matrix representation B is that

$$\eta_\beta = B\xi_\beta.$$

Substitution from (27) then gives

$$P\eta_\alpha = BP\xi_\alpha, \quad \eta_\alpha = P^{-1}BP\xi_\alpha.$$

The last equation shows that $P^{-1}BP$ represents T relative to the basis of α_i, whereas this matrix representation is unique and is A. This gives (26).

EXERCISES

★1. Let T be a linear transformation of $V_n(\mathfrak{F})$, and let A be a matrix representing T. Prove that A is similar to a diagonal matrix if and only if $V_n(\mathfrak{F})$ has a basis $\alpha_1, \ldots, \alpha_n$ such that $T'(\alpha_i) = c_i\alpha_i$, $i = 1, \ldots, n$, the c_i being scalars.

2. Let A be an idempotent matrix and let T be a linear transformation represented by A. Show that T is of the type discussed in Exercise 8, Section 10-3, and prove that A is similar to the diagonal matrix

$$\begin{bmatrix} I_r & 0 \\ 0 & 0 \end{bmatrix}.$$

(Hint: Use bases of M and N in Exercise 8, Section 10-3.) Finally, prove that two $n \times n$ idempotent matrices over $\mathfrak{F}$ are similar if and only if they have the same rank.

10–6 Orthogonal complement of a subspace. If M is an arbitrary subspace of $V_n(\mathfrak{F})$, where $\mathfrak{F} = \mathfrak{R}$ or $\mathfrak{C}$, the totality of vectors in $V_n(\mathfrak{F})$ orthogonal to every vector in M is a set $\mathfrak{O}(M)$ called the *orthogonal complement* of M.

THEOREM 10–8. *Let M be a subspace of $V_n(\mathfrak{F})$, $\mathfrak{F} = \mathfrak{R}$ or $\mathfrak{C}$. The orthogonal complement $\mathfrak{O}(M)$ of M is also a subspace of $V_n(\mathfrak{F})$, and every vector in $V_n(\mathfrak{F})$ is expressible uniquely in the form $\mu + \nu$, μ in M, ν in $\mathfrak{O}(M)$.*

Let $\alpha_1, \ldots, \alpha_m$ form an orthonormal (or normal unitary) basis for M. Then (Theorem 9–8 or Theorem 9–15) there are further vectors α_i ($i = m + 1, \ldots, n$) such that all n vectors α_i form an orthonormal (or normal unitary) basis for $V_n(\mathfrak{F})$. The space S spanned by the new vectors

$$(28) \qquad \alpha_{m+1}, \ldots, \alpha_n$$

may be shown to be $\mathfrak{O}(M)$, and every assertion in the theorem is a clear consequence of the equality $S = \mathfrak{O}(M)$. To verify this equality one may observe first that $S \leqq \mathfrak{O}(M)$. (Why?) For the reverse inclusion, let

$$\eta = b_1\alpha_1 + \cdots + b_n\alpha_n$$

be any vector of $\mathfrak{O}(M)$. Then each of $\alpha_1, \ldots, \alpha_m$ is orthogonal to η:

$$0 = \bar{\alpha}_i'\eta = \sum_j \bar{\alpha}_i'b_j\alpha_j = b_i. \qquad (i = 1, \ldots, m)$$

Thus η is in the space spanned by (28), so that $\mathfrak{O}(M) \leqq S$ and the theorem is established.

The sum of two subspaces of $V_n(\mathfrak{F})$ was defined in Section 2–8. Part of Theorem 10–8 may be summarized in the statement that $V_n(\mathfrak{F})$ is the sum

$$(29) \qquad V_n(\mathfrak{F}) = M + \mathfrak{O}(M).$$

Since every vector ξ in $V_n(\mathfrak{F})$ determines unique vectors ξ_M in M and ν in $\mathfrak{O}(M)$ such that $\xi = \xi_M + \nu$, the correspondence

$$(30) \qquad \xi \to \xi_M = T(\xi)$$

is well defined by the subspace M, and maps each vector in $V_n(\mathfrak{F})$ into a vector in M. The vector $T(\xi)$ defined here is called the *perpendicular projection* of ξ on M. The mapping (30) carrying each ξ into its perpendicular projection on a fixed subspace M is called a *perpendicular projection on M*.

EXERCISES

1. Prove the inclusion $S \leqq \Theta(M)$ occurring in the proof of Theorem 10–8.

2. Prove directly from the definition of $\Theta(M)$ that it is a subspace.

3. Prove that $\Theta[\Theta(M)] = M$.

★4. Let T be a perpendicular projection on a subspace M. Prove that T is a linear transformation and that $T[T(\xi)] = T(\xi)$ for every vector ξ.

5. Show that T in Exercise 4 has a matrix representation diag $(I_m, 0)$ where m is the dimension of M.

★6. A space S is the *direct sum* of two spaces U and V contained in S in case $S = U + V$ and the intersection (see Section 2–8) of U and V is zero. We write $S = U \oplus V$ to indicate that S is the direct sum of U and V. Show that the property $S = U \oplus V$ is equivalent to each of the properties (a) and (b):

(a) $S = U + V$, and the dimension of S is the sum of the dimensions of U and V.

(b) $S = U + V$, and the expression of any vector ξ of S in the form $\xi = \alpha + \beta$, α in U, β in V, is unique.

★7. Let T be a linear transformation of $V_n(\mathfrak{F})$. Prove that T is represented by a direct sum diag (A, B) relative to a suitable basis if and only if $V_n(\mathfrak{F})$ is a direct sum $U \oplus V$ such that T maps U into U and V into V.

10–7 Orthogonal transformations. In the real plane one may use oblique coordinate systems (see Figure 10–4) as well as the more familiar rectangular coordinate systems. In terms of oblique coordinates the distance of the point (X, Y) from the origin is

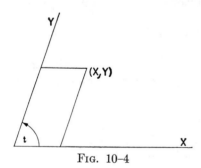

$$(X^2 + 2XY \cos t + Y^2)^{\frac{1}{2}},$$

whereas in terms of rectangular coordinates (x,y) the same distance is given by the simpler formula

$$(31) \qquad (x^2 + y^2)^{\frac{1}{2}}.$$

FIG. 10–4

The length of $\xi = \text{col } (x_1, \ldots, x_n)$ in $V_n(\mathfrak{R})$ is given by a formula which generalizes (31):

$$(32) \qquad (x_1^2 + \cdots + x_n^2)^{\frac{1}{2}}.$$

The x_i here are coordinates of ξ relative to a particular orthonormal basis, namely, the unit vectors. We shall see that a formula just like (32) is valid if instead of the x_i we use coordinates of ξ relative to an arbitrary orthonormal basis

$$\alpha_1, \ldots, \alpha_n$$

of $V_n(\mathfrak{R})$. The matrix M having α_i as its ith column, $i = 1, \ldots, n$ is orthogonal: $M'M = I$. If

(33) $\xi_\alpha = \text{col }(a_1, \ldots, a_n), \quad \eta_\alpha = \text{col }(b_1, \ldots, b_n),$

we have

$$\xi = M\xi_\alpha, \quad \eta = M\eta_\alpha,$$
$$\xi'\eta = \xi'_\alpha M'M\eta_\alpha = \xi'_\alpha \eta_\alpha,$$

(34) $\xi'\eta = a_1b_1 + \cdots + a_nb_n.$

This formula for the inner product of ξ and η has the same appearance as the defining formula in which we used coordinates relative to the unit vectors.

Hence we conclude that in computing the inner product of two vectors of $V_n(\mathfrak{R})$ we may use coordinates relative to an arbitrary orthonormal basis. Thus, vectors η and ξ with coordinates (33) are orthogonal if and only if their inner product (34) is zero. Also, the length of ξ is

$$(a_1^2 + \cdots + a_n^2)^{\frac{1}{2}}.$$

By virtue of these properties the orthonormal bases of $V_n(\mathfrak{R})$ effectively generalize the concept of rectangular coordinate systems in real space of two and three dimensions.

LEMMA 10–1. *Let $\alpha_1, \ldots, \alpha_n$ be any orthonormal basis of $V_n(\mathfrak{R})$, relative to which a linear transformation T is represented by a matrix A. Then T is orthogonal if and only if A is orthogonal.*

Let $\xi_\alpha = \text{col }(a_1, \ldots, a_n)$. The vector $T(\xi)$ has coordinates $A\xi_\alpha$ which, by the observations above, may be used in computing the length-squared of $T(\xi)$:

(35) $T(\xi)'T(\xi) = \xi'_\alpha A'A\xi_\alpha.$

Now if A is orthogonal, $A'A = I$ so that

(36) $T(\xi)'T(\xi) = \xi'_\alpha \xi_\alpha.$

Thus T preserves lengths, and this is the defining property of an orthogonal transformation. Conversely, if T is orthogonal both (36) and (35) are valid and

$$(a_1, \ldots, a_n)A'A \text{ col }(a_1, \ldots, a_n) = a_1^2 + \cdots + a_n^2$$

for all scalars $a_1, \ldots, a_n$. Then by Theorem 5–3, $A'A = I$.

THEOREM 10–9. *Let T be an orthogonal transformation of $V_n(\mathfrak{R})$, $n = 2$ or 3. Then there is a rectangular coordinate system such that T is either a rotation of coordinate axes, a reflection in a coordinate plane or axis, or a succession of two such mappings.*

First, let n be arbitrary. The matrix A representing T relative to any orthonormal basis is orthogonal, hence (Theorem 9–24) its characteristic roots have absolute value equal to unity. Each real root then is 1 or -1, and the imaginary roots are pairs $\cos t \pm i \sin t$. Since an orthogonal matrix is real and normal, Theorem 9–27 applies so that A is orthogonally similar to a matrix of the type

$$B = \operatorname{diag}\,(-I_q,\, I_p,\, C_1,\, \ldots,\, C_m),$$

$$C_j = \begin{bmatrix} \cos t_j & -\sin t_j \\ \sin t_j & \cos t_j \end{bmatrix}. \qquad (j = 1, \ldots, m)$$

If $q > 1$, 2×2 blocks taken from $-I_q$ may be regarded as follows:

$$\begin{bmatrix} -1 & 0 \\ 0 & -1 \end{bmatrix} = \begin{bmatrix} \cos \pi & -\sin \pi \\ \sin \pi & \cos \pi \end{bmatrix}.$$

Since each such block may be regarded as a C_j, it may be assumed hereafter that either $-I_q$ is absent entirely or that it is 1×1. The matrix B is clearly a product

(37) $$B = ED_1 \cdots D_m = D_m \cdots D_1 E,$$

where E is either I_n or obtainable from I_n by changing the first diagonal element to -1, and D_j is obtainable from I_n by substituting C_j for a 2×2 diagonal block in a suitable position, the same position as that occupied by C_j as a block in B.

The similarity of B and A implies that B represents the given transformation T relative to a new basis $\beta_1, \ldots, \beta_n$. Since B is orthogonally similar to A, the new basis, like the old, is orthonormal. (Why?) In terms of coordinates relative to this new basis T is given by the formula

$$T(\xi)_\beta = B\xi_\beta = D_m \cdots D_1 E\xi_\beta.$$

If R_j and S are the transformations

$$R_j(\xi)_\beta = D_j\xi_\beta, \qquad (j = 1, \ldots, m)$$
$$S(\xi)_\beta = E\xi_\beta,$$

we see that T is accomplished by performing in succession S, R_1, $\ldots, R_m$.

When $n = 3$, B is either E or ED, where

$$D = \begin{bmatrix} 1 & 0 & 0 \\ 0 & \cos t & -\sin t \\ 0 & \sin t & \cos t \end{bmatrix}, \quad E = \begin{bmatrix} \pm 1 & 0 & 0 \\ 0 & 1 & 0 \\ 0 & 0 & 1 \end{bmatrix}.$$

If the new orthonormal basis is interpreted as an xyz rectangular co-ordinate system, E represents either the identity transformation (carrying each point into itself) or the transformation carrying each point (x, y, z) into (x', y', z'), where

$$x' = -x, \quad y' = y, \quad z' = z.$$

This is a reflection in the yz plane. Finally, D represents the transformation

$$\begin{aligned} x' &= x, \\ y' &= y \cos t - z \sin t, \\ z' &= y \sin t + z \cos t, \end{aligned}$$

which is a rotation about the x axis. The case $n = 2$ may be analyzed similarly. Theorem 10-9 is commonly stated without any restriction on n, rotations and reflections in $V_n(\Re)$ being suitably defined.

EXERCISES

1. Consider the locus in real two-dimensional space of the equation $X'AX = 1$, where $X'AX$ is a quadratic form over $\Re$ and $X' = (x, y)$. Let r and s denote the characteristic roots of the real, symmetric matrix A. Show that the locus is an ellipse if both r and s are positive, a hyperbola if they have opposite signs, and nonexistent if both r and s are negative; also, that there is no locus if both roots are zero or if one root is zero and the other negative; and that the locus is two straight lines if one root is zero and one is positive.

2. Referring to Exercise 1, show that the locus is nonexistent if the index of A is zero, and existent if the index is positive. In the latter case show that the locus is a nondegenerate conic if A is nonsingular, and a degenerate conic if A is singular.

3. Let T be an orthogonal transformation of $V_2(\Re)$ with matrix representation A. Show that T is a reflection in a line if $|A| = -1$, and a rotation if $|A| = 1$.

4. State and prove the analogue of Lemma 10-1 for unitary transformations of $V_n(\mathbb{C})$.

5. Define rotations and reflections in $V_n(\Re)$ so that Theorem 10-9 is valid for arbitrary n.

10–8 Operations on linear transformations. In Section 10–7 we considered performing successively a number of reflections and rotations. Suppose now that S and T are any two linear transformations of $V_n(\mathfrak{F})$. If we perform first T, then S, that is,

$$\xi \to T(\xi) \to S[T(\xi)] = ST(\xi),$$

each vector ξ is mapped into a vector which is commonly indicated as $ST(\xi)$. The correspondence

(38) $$\xi \to ST(\xi)$$

may be regarded as a single entity. It is actually a linear transformation of $V_n(\mathfrak{F})$:

$$\begin{aligned}
ST(c_1\xi_1 + c_2\xi_2) &= S[T(c_1\xi_1 + c_2\xi_2)] \\
&= S[c_1T(\xi_1) + c_2T(\xi_2)] \\
&= c_1S[T(\xi_1)] + c_2S[T(\xi_2)] \\
&= c_1ST(\xi_1) + c_2ST(\xi_2).
\end{aligned}$$

Thus (38) is a linear transformation. It is called the *product* of S and T.

Suppose that S and T are represented by matrices S_0 and T_0, respectively, relative to some fixed basis. Then $T(\xi)$ has coordinates $T_0\xi_\alpha$, so that $\eta = ST(\xi)$ may be calculated in terms of coordinates as follows:

(39) $$\eta_\alpha = S_0 T_0 \xi_\alpha.$$

This shows that the matrix representing ST is $S_0 T_0$.

The *sum* $S + T$ of linear transformations S and T is defined to be the correspondence

(40) $$\xi \to S(\xi) + T(\xi).$$

The sum $S(\xi) + T(\xi)$ may be indicated as

$$(S + T)\xi.$$

The correspondence (40) is linear, since

$$\begin{aligned}
S(c_1\xi_1 &+ c_2\xi_2) + T(c_1\xi_1 + c_2\xi_2) \\
&= c_1S(\xi_1) + c_2S(\xi_2) + c_1T(\xi_1) + c_2T(\xi_2) \\
&= c_1[S(\xi_1) + T(\xi_1)] + c_2[S(\xi_2) + T(\xi_2)] \\
&= c_1[(S + T)\xi_1] + c_2[(S + T)\xi_2].
\end{aligned}$$

In terms of coordinates $\eta = (S + T)\xi$ may be calculated as follows:

(41) $$\eta_\alpha = S_0\xi_\alpha + T_0\xi_\alpha = (S_0 + T_0)\xi_\alpha.$$

This shows that the linear transformation $S + T$ is represented by $S_0 + T_0$.

Scalar multiplication also arises for linear transformations. If k is a scalar, the correspondence

$$(42) \qquad\qquad \xi \to k\xi$$

is easily seen to be a linear transformation S_k. Moreover, relative to any basis whatsoever, S_k is represented by the matrix $S_0 = kI$. By the fact proved in (39) the product $S_k T$ is represented by $kI \cdot T_0 = kT_0$. These ideas motivate the following definition of the product of a scalar k by a linear transformation T: The product kT is defined to be the product $S_k T$, where S_k is the linear transformation (42). Thus the transformation kT is effected on a given vector ξ by first finding $T(\xi)$, then multiplying this result by the scalar k. As we have seen, kT is represented by the matrix kT_0.

In summary, the set $L_n(\mathfrak{F})$ of all linear transformations on $V_n(\mathfrak{F})$ is subject to three operations called addition, multiplication, and scalar multiplication. The set $\mathfrak{F}_n$ of all $n \times n$ matrices over $\mathfrak{F}$ is subject to three like-named operations. Once a basis of $V_n(\mathfrak{F})$ is selected, each member S of $L_n(\mathfrak{F})$ corresponds to a unique member S_0 of $\mathfrak{F}_n$, S_0 being the matrix representing S relative to the chosen basis; each matrix S_0 in $\mathfrak{F}_n$ represents a unique S in $L_n(\mathfrak{F})$. Moreover, the matrices corresponding to the linear transformations ST, $S + T$, and kT are, respectively,

$$S_0 T_0, \quad S_0 + T_0, \quad kT_0,$$

where S_0 and T_0 are the matrices corresponding to S and T. Thus computations involving linear transformations may be carried out effectively by use of the corresponding matrices, and conversely computations involving matrices may be interpreted by means of linear transformations. For these reasons the study of $L_n(\mathfrak{F})$ on the one hand and of $\mathfrak{F}_n$ on the other amount to two ways of looking at the same subject. While more abstract, the linear transformations can in some ways be handled with greater facility than their more concrete counterparts, matrices. The reader who is interested in this approach may wish to read References 9 and 14.

EXERCISES

1. Prove that (42) is a linear transformation and that it is represented relative to any basis by the matrix kI.

2. What is the meaning of $(k_1 S_1 + \cdots + k_v S_t)\xi$, where ξ is a vector, the S_i are linear transformations, and the k_i are scalars?

3. Let a matrix A represent a linear transformation T. Prove that A is nonsingular if and only if T is one-to-one (that is, maps distinct vectors into distinct vectors).

10–9 Summary. In the main, this book has been devoted to the study of seven RST relations on certain classes of matrices. These relations and some of the vital facts pertaining thereto may be summarized as follows.

I. **Equivalence over $\mathfrak{F}$:** $B = PAQ$.

P and Q nonsingular.

A and B rectangular.

Arbitrary scalar field $\mathfrak{F}$.

II. **Equivalence over $\mathfrak{F}[x]$:** $B = PAQ$.

P and Q nonsingular matrices over $\mathfrak{F}[x]$ having determinants in $\mathfrak{F}$, hence having inverses with elements in $\mathfrak{F}[x]$.

A and B rectangular.

$\mathfrak{F}$ is an arbitrary field, and the scalars are polynomials in one variable over $\mathfrak{F}$.

III. **Similarity:** $B = P^{-1}AP$.

P nonsingular.

Arbitrary scalar field $\mathfrak{F}$.

IV. **Congruence:** $B = P'AP$.

P nonsingular.

Results obtained only when scalar field $\mathfrak{F}$ obeys $1 + 1 \neq 0$, and for the major results only when $\mathfrak{F} = \mathfrak{R}$.

V. **Hermitian congruence:** $B = \bar{P}'AP$.

P nonsingular.

Field of scalars is the field $\mathfrak{C}$ of all complex numbers.

VI. **Orthogonal similarity:** $B = P^{-1}AP$.

$P^{-1} = P'$.

Field of scalars is the field $\mathfrak{R}$ of all real numbers.

VII. **Unitary similarity:** $B = P^{-1}AP$.

$P^{-1} = \bar{P}'$.

Field of scalars is $\mathfrak{C}$.

All of the relations except the first two require the related matrices A and B to be square. Otherwise the products defining the relations are impossible.

For each of the seven relations there are two major problems:

(1) obtaining a set of canonical matrices; (2) finding a necessary and sufficient condition for two matrices to obey the specified relation (i.e., to be similar or equivalent or congruent, etc.). These problems were solved in full generality for the first three relations, but only for restricted classes of matrices in the last four relations.

The results for similarity (III) are interesting in that several canonical sets are available. One canonical matrix similar to a given matrix A is the direct sum B of the hypercompanion matrices of the elementary divisors of A. If the matrix A is fixed but the scalar field is progressively enlarged, this canonical matrix B changes; if the field is suitably enlarged, B then becomes a classical canonical matrix.

Notable results for congruence (IV) were obtained only when attention was confined to the class of symmetric matrices over the real number field, or the class of skew matrices over a general field in which $1 + 1 \neq 0$. Both of the major problems above were solved for these cases. The theory of Hermitian congruence (V) was worked out primarily for the case of Hermitian matrices, the results being parallel to those for congruence of real symmetric matrices.

Orthogonal matrices, and real symmetric and real skew matrices are special cases of real normal matrices. The two major problems were solved for the orthogonal similarity (VI) of real normal matrices. The complex analogue is concerned with the unitary similarity (VII) of general normal matrices; for this case the two major problems were solved, thereby covering the special cases of matrices which are Hermitian or unitary.

Having thus reviewed the seven major relations from the viewpoints of definitions, the generality of the scalar system, and the particular classes of matrices for which canonical sets were developed, we shall now review the concrete situations in which the various relations appear. Nonsingular linear substitutions in bilinear forms, quadratic forms, and Hermitian forms give rise to equivalence, congruence, and Hermitian congruence, respectively. For the last two types of form, if the substitution is required to be length-preserving, one is led to orthogonal and unitary similarity. These two relations play roles also in the simultaneous reduction of a pair of forms. Equivalence over $\mathfrak{F}[x]$ is primarily a tool for the study of similarity. The latter relation occurs in the determination of all matrices representing a common linear transformation, and also in concrete form in connection with solutions of systems of differential equations.

REFERENCES

1. Aitken, A. C., *Determinants and Matrices*. Edinburgh: Oliver and Boyd, 1948.

2. Albert, A. A., *Introduction to Algebraic Theories*. Chicago: The University of Chicago Press, 1941.

3. Albert, A. A., *Modern Higher Algebra*. Chicago: The University of Chicago Press, 1937.

4. Birkhoff, G., and MacLane, S., *A Survey of Modern Algebra*. New York: Macmillan, 1941.

5. Bocher, M., *Introduction to Higher Algebra*. New York: Macmillan, 1907.

6. Dickson, L. E., *Modern Algebraic Theories*. Chicago: Sanborn, 1926.

7. Ferrar, W. L., *Algebra*. Oxford: Oxford University Press, 1941.

8. Frazer, R. A., Duncan, W. J., and Collar, A. R., *Elementary Matrices and Some Applications to Dynamics and Differential Equations*. Cambridge: The University Press, 1938.

9. Halmos, P. R., *Finite Dimensional Vector Spaces*. Princeton: Princeton University Press, 1942.

10. MacDuffee, C. C., *An Introduction to Abstract Algebra*. New York: Wiley, 1940.

11. MacDuffee, C. C., *The Theory of Matrices*. New York: Chelsea, 1946.

12. MacDuffee, C. C., *Vectors and Matrices (Carus Math. Monograph, No. 7)*. The Mathematical Association of America, 1943.

13. Michal, A. D., *Matrix and Tensor Calculus*. New York: Wiley, 1947.

14. Schreier, O., and Sperner, E., *Vorlesungen über Matrizen*. Leipzig: Teubner, 1932.

15. Turnbull, H. W., and Aitken, A. C., *An Introduction to the Theory of Canonical Matrices*. London: Blackie, 1948.

16. Uspensky, J. V., *Theory of Equations*. New York: McGraw-Hill, 1948.

17. Waerden, B. L. van der, *Modern Algebra*, vol. 1 (translated from German). New York: Ungar, 1949.

18. Wedderburn, J. H. M., *Lectures on Matrices*. New York: American Mathematical Society (*Colloquium Publications*, vol. 17), 1934.

For information on special topics not treated in the present text or in the references cited above, the reader would do well to acquaint himself with *Mathematical Reviews*. This is a monthly periodical which gives short summaries of practically all research articles appearing in current mathematical journals. It is extensively indexed to permit easy selection of the reviews of all articles on a particular subject.

INDEX

adjoint, 75
alternate matrix, 95
associative law, 7

basis, 28
belong to, 18
bilinear form, 65
  matrix of, 66
  rank of, 66
block multiplication, 15

canonical set, 58
Cayley-Hamilton theorem, 136
characteristic
  equation, 136
  function, 136
  matrix, 136
  polynomial, 136
  roots, 166, 169
  vector, 170
class, 18
classical canonical form, 163
closed, 18, 25
cofactor, 75
column
  operations, 49
  rank, 49
  space, 49
commutative law, 7
commutes with, 7
companion matrix, 148
components, 24
congruent automorphs, 87 (Ex. 7),
  103
congruent matrices, 86
conjugate, 97
conjunctive matrices, 99
constant polynomials, 109
contained in, 18
contains, 18

coordinates, 1, 24, 214
Cramer's rule, 83

decomposable matrix, 154
degree of a polynomial, 109
degree of a matric polynomial, 131
determinant, 69
diagonable matrix, 177
diagonal
  elements, 6
  matrix, 9
  of a matrix, 6
dimension, 32
direct sum
  of matrices, 16
  of subspaces, 226 (Ex. 6)
distributive law, 10
divides, 109
division algorithm, 110, 134
divisor of zero, 20

eigenvalues, 169
eigenvector, 170
elementary
  column operations, 49
  divisors, 157
  matrix, 40, 123
  operations, 51, 123
  row operations, 38
elements, 1
entries, 1
equivalent
  matrices, 51, 123
  systems of equations, 36, 139
expansion of a determinant, 77

factor, 109
factorization, 110
  trivial, 110
field, 17

greatest common divisor, 113, 120

A CATALOG OF SELECTED
DOVER BOOKS
IN SCIENCE AND MATHEMATICS

A CATALOG OF SELECTED
DOVER BOOKS
IN SCIENCE AND MATHEMATICS

QUALITATIVE THEORY OF DIFFERENTIAL EQUATIONS, V.V. Nemytskii and V.V. Stepanov. Classic graduate-level text by two prominent Soviet mathematicians covers classical differential equations as well as topological dynamics and erqodic theory. Bibliographies. 523pp. 5⅜ × 8½. 65954-2 Pa. $10.95

MATRICES AND LINEAR ALGEBRA, Hans Schneider and George Phillip Barker. Basic textbook covers theory of matrices and its applications to systems of linear equations and related topics such as determinants, eigenvalues and differential equations. Numerous exercises. 432pp. 5⅜ × 8½. 66014-1 Pa. $8.95

QUANTUM THEORY, David Bohm. This advanced undergraduate-level text presents the quantum theory in terms of qualitative and imaginative concepts, followed by specific applications worked out in mathematical detail. Preface. Index. 655pp. 5⅜ × 8½. 65969-0 Pa. $10.95

ATOMIC PHYSICS (8th edition), Max Born. Nobel laureate's lucid treatment of kinetic theory of gases, elementary particles, nuclear atom, wave-corpuscles, atomic structure and spectral lines, much more. Over 40 appendices, bibliography. 495pp. 5⅜ × 8½. 65984-4 Pa. $11.95

ELECTRONIC STRUCTURE AND THE PROPERTIES OF SOLIDS: The Physics of the Chemical Bond, Walter A. Harrison. Innovative text offers basic understanding of the electronic structure of covalent and ionic solids, simple metals, transition metals and their compounds. Problems. 1980 edition. 582pp. 6⅛ × 9¼. 66021-4 Pa. $14.95

BOUNDARY VALUE PROBLEMS OF HEAT CONDUCTION, M. Necati Özisik. Systematic, comprehensive treatment of modern mathematical methods of solving problems in heat conduction and diffusion. Numerous examples and problems. Selected references. Appendices. 505pp. 5⅜ × 8½. 65990-9 Pa. $11.95

A SHORT HISTORY OF CHEMISTRY (3rd edition), J.R. Partington. Classic exposition explores origins of chemistry, alchemy, early medical chemistry, nature of atmosphere, theory of valency, laws and structure of atomic theory, much more. 428pp. 5⅜ × 8½. (Available in U.S. only) 65977-1 Pa. $10.95

A HISTORY OF ASTRONOMY, A. Pannekoek. Well-balanced, carefully reasoned study covers such topics as Ptolemaic theory, work of Copernicus, Kepler, Newton, Eddington's work on stars, much more. Illustrated. References. 521pp. 5⅜ × 8½. 65994-1 Pa. $11.95

PRINCIPLES OF METEOROLOGICAL ANALYSIS, Walter J. Saucier. Highly respected, abundantly illustrated classic reviews atmospheric variables, hydrostatics, static stability, various analyses (scalar, cross-section, isobaric, isentropic, more). For intermediate meteorology students. 454pp. 6⅛ × 9¼. 65979-8 Pa. $12.95

CATALOG OF DOVER BOOKS

CHALLENGING MATHEMATICAL PROBLEMS WITH ELEMENTARY SOLUTIONS, A.M. Yaglom and I.M. Yaglom. Over 170 challenging problems on probability theory, combinatorial analysis, points and lines, topology, convex polygons, many other topics. Solutions. Total of 445pp. 5⅜ × 8½. Two-vol. set.
Vol. I 65536-9 Pa. $5.95
Vol. II 65537-7 Pa. $5.95

FIFTY CHALLENGING PROBLEMS IN PROBABILITY WITH SOLUTIONS, Frederick Mosteller. Remarkable puzzlers, graded in difficulty, illustrate elementary and advanced aspects of probability. Detailed solutions. 88pp. 5⅜ × 8½.
65355-2 Pa. $3.95

EXPERIMENTS IN TOPOLOGY, Stephen Barr. Classic, lively explanation of one of the byways of mathematics. Klein bottles, Moebius strips, projective planes, map coloring, problem of the Koenigsberg bridges, much more, described with clarity and wit. 43 figures. 210pp. 5⅜ × 8½. 25933-1 Pa. $4.95

RELATIVITY IN ILLUSTRATIONS, Jacob T. Schwartz. Clear non-technical treatment makes relativity more accessible than ever before. Over 60 drawings illustrate concepts more clearly than text alone. Only high school geometry needed. Bibliography. 128pp. 6⅛ × 9¼. 25965-X Pa. $5.95

AN INTRODUCTION TO ORDINARY DIFFERENTIAL EQUATIONS, Earl A. Coddington. A thorough and systematic first course in elementary differential equations for undergraduates in mathematics and science, with many exercises and problems (with answers). Index. 304pp. 5⅜ × 8¼. 65942-9 Pa. $7.95

FOURIER SERIES AND ORTHOGONAL FUNCTIONS, Harry F. Davis. An incisive text combining theory and practical example to introduce Fourier series, orthogonal functions and applications of the Fourier method to boundary-value problems. 570 exercises. Answers and notes. 416pp. 5⅜ × 8½. 65973-9 Pa. $8.95

THE THOERY OF BRANCHING PROCESSES, Theodore E. Harris. First systematic, comprehensive treatment of branching (i.e. multiplicative) processes and their applications. Galton-Watson model, Markov branching processes, electron-photon cascade, many other topics. Rigorous proofs. Bibliography. 240pp. 5⅜ × 8½. 65952-6 Pa. $6.95

AN INTRODUCTION TO ALGEBRAIC STRUCTURES, Joseph Landin. Superb self-contained text covers "abstract algebra": sets and numbers, theory of groups, theory of rings, much more. Numerous well-chosen examples, exercises. 247pp. 5⅜ × 8½. 65940-2 Pa. $6.95

GAMES AND DECISIONS: Introduction and Critical Survey, R. Duncan Luce and Howard Raiffa. Superb non-technical introduction to game theory, primarily applied to social sciences. Utility theory, zero-sum games, n-person games, decision-making, much more. Bibliography. 509pp. 5⅜ × 8½. 65943-7 Pa. $10.95

Prices subject to change without notice.

Available at your book dealer or write for free Mathematics and Science Catalog to Dept. GI, Dover Publications, Inc., 31 East 2nd St., Mineola, N.Y. 11501. Dover publishes more than 175 books each year on science, elementary and advanced mathematics, biology, music, art, literary history, social sciences and other areas.